Praise for

"Who knew subject matter so (literally) uncomfortable could be so much damn fun? Shaughnessy Bishop-Stall is the perfect endearingly flawed and funny narrator to take us on this wild, worldwide adventure into the history of our painful mornings after. Whether he's piloting a fighter plane in Vegas, chatting with a blacksmith in Devon, cheating death in the desert and the alps, or attempting twelve pints in twelve pubs, his daring, wit, and insight never disappoint—all with, it would seem, a blazing hangover. Part science, part folklore, part string of the author's very bad ideas with good intentions, *Hungover* is a highly knowledgeable and ridiculously enjoyable ride."

—Stacey May Fowles, author of *Baseball Life Advice*

"Shaughnessy Bishop-Stall has risked life and liver to write this book, a perilous trip into many mornings after—historical, cinematic, literary, and of course his own. That's the entertaining part. As to whether Bishop-Stall has, in fact, invented a hangover remedy that actually works? I live in hope."

—Adam Rogers, author of *Proof: The Science of Booze*

"It takes a writer as skilled as Shaughnessy Bishop-Stall to write a rip-roaring adventure story about the morning after. Thoroughly researched, rich in history and humor, against all odds, *Hungover* makes you wish you were there."

—Tabatha Southey, author of *The Deep Cold River Story*

"Shaughnessy Bishop-Stall invests health, wealth, and well-being in a wild Dionysian quest for a viable hangover cure. In the end he gets more than one, and we do, too."

—Linden MacIntyre, Scotiabank Giller Prize–winning author of *The Bishop's Man*

Praise for *Down to This*

"Nothing short of a masterpiece." —*National Post* (Canada)

"Some writers go to great lengths to write a book. They climb Mount Everest, follow armies into war zones, go undercover with a professional sports team, or travel around the world on a motorbike. . . . Shaughnessy Bishop-Stall has more guts than any of those writers." —*Edmonton Journal*

"Intensely perceptive, Bishop-Stall tumbles heartbreak with hilarity, outrageous despair with shimmering hope." —*Calgary Herald*

PENGUIN BOOKS

HUNGOVER

Shaughnessy Bishop-Stall's first book, *Down to This: Squalor and Splendour in a Big-City Shantytown*—about a year he spent living with the homeless—was shortlisted for several prestigious awards, none of which it won. His first novel, *Ghosted*, about a guy who becomes a professional ghostwriter of suicide letters, was nominated for the Amazon First Novel Award, which it also lost. His work has appeared in dozens of magazines—most of which no longer exist. He played the role of Jason, a well-dressed, bad-mannered journalist, on CBC's *The Newsroom*—in what turned out to be its final season. He used to own a bar, called The Lowdown, but that didn't work out either. He is not very good at social media—but he also has trouble letting things go, so you might want to check for updates on his quest and improvements on the cure at hungoverlowdown.com.

SHAUGHNESSY BISHOP-STALL

HUNG

OVER

THE MORNING AFTER AND
ONE MAN'S QUEST FOR THE CURE

PENGUIN BOOKS

PENGUIN BOOKS
An imprint of Penguin Random House LLC
penguinrandomhouse.com

LIBRARY OF CONGRESS CATALOGING-IN-PUBLICATION DATA
Names: Bishop-Stall, Shaughnessy, author.
Title: Hungover : the morning after and one man's quest for the cure /
Shaughnessy Bishop-Stall.
Description: New York, NY : Penguin Books, [2018]
Identifiers: LCCN 2018021163 (print) | LCCN 2018021546 (ebook) |
ISBN 9780698178939 (ebook) | ISBN 9780143126706 (paperback)
Subjects: LCSH: Detoxification (Substance abuse treatment)—Popular works. |
Alcohol—Physiological effect—Popular works. | Hangover cures—Popular works. |
BISAC: COOKING / Essays. | COOKING / Beverages / Wine & Spirits. |
SCIENCE / Life Sciences / General.
Classification: LCC RC565 (ebook) | LCC RC565 .B52 2018 (print) |
DDC 362.29/18—dc23
LC record available at https://lccn.loc.gov/2018021163

Printed in the United States of America
3   5   7   9   10   8   6   4   2

Penguin is committed to publishing works of quality and integrity.
In that spirit, we are proud to offer this book to our readers; however,
the story, the experiences, and the words are the author's alone.

For Brandy Bob Stall,
Who lived so fast,
He soared right past his day of dying,
And still has never been hungover,
though not, of course, for lack of trying.

Spoiler, Disclaimer and Full Disclosure from the Author

This book took almost a decade to write. And as of the time of this note, I am still alive. That's the spoiler.

As a disclaimer: this subject proved to be far richer than I first imagined. Though my original intention was to make it an entirely global venture, the narrative takes place mostly in what we call the West. I hope, one day—after much recuperation—to delve more deeply into Russia, Asia and Africa, and farther south into South America.

In terms of full disclosure: over the past many years I have traveled to too many cities in too many countries and drunk far too much of everything—with barkeeps, businesspeople and brewers, winemakers, winos and whiners, distillers, doctors and druids, as well as some people I probably shouldn't have. I have tried every tincture, tonic, powder, pill, placebo, root, leaf, bark, chemical and therapeutic process I could legally test, and then some others. And although everything on these pages did take place and I've done my damnedest with the fact-checking, the order in which some events appear is not always chronological. No matter how I tried, however, the morning after did still follow the night before.

Contents

Preface: A Few Words About a Few Words.............................XIII

Welcome to Your Hangover 1

PART ONE: WHAT HAPPENS IN VEGAS..............................7
First Interlude: A Drink Before the War 32

PART TWO: WHAT HAPPENS *ABOVE* VEGAS 37
Second Interlude: Plenty of Aversion; A Version of Pliny..........61

PART THREE: THE HAIR THAT WAGS THE DOG......................65
Third Interlude: And Up She Rises 108

PART FOUR: A MAD HATTER IN MIDDLE EARTH...................115
Fourth Interlude: Werewolves of London........................140

PART FIVE: TWELVE PINTS IN TWELVE PUBS....................145
Fifth Interlude: The Withnail Awards; A Press Release..........176

PART SIX: THE HUNGOVER GAMES 179
Sixth Interlude: A Roots of Remedy Roundup 206

PART SEVEN: THE FUTURE'S SO BRIGHT 211
Seventh Interlude: Killer Parties 235

PART EIGHT: THE TIGER ON THE ROOF 239

Eighth Interlude: I Woke Up This Morning 256

PART NINE: BEYOND THE VOLCANOES 261

Ninth Interlude: Aspirin or Sorrow 290

PART TEN: WHEN LIZARDS DRINK FROM YOUR EYES 295

Tenth Interlude: The Hangover Writer 317

PART ELEVEN: AFTER THE FLOOD 323

For the Love of Hangovers: A Kind of Conclusion 355

Acknowledgments 371

Permissions 377

Notes on Sources 379

Bibliography 391

A FEW WORDS ABOUT A FEW WORDS

A title is the start of any story that happens to have a title. And this one has already caused some controversy, at least with my editor, who is pressing for a hyphen (*Hung-over*), and my father, who is adamant it be two words (*Hung Over*). But in my opinion, one of them drinks a helluva lot, and the other not quite enough. And anyway, it is my book, so we'll be going with *hungover*—along with *helluva*, *alright* and *goddamn*.

Hungover is an adjective, derived from the noun *hangover*, not to be confused with *drunk*—a difference well explained in Richard Linklater's 2003 film *School of Rock*:

> DEWEY FINN (Jack Black): *Okay. Here's the deal. I have a hangover. Who knows what that means?*
> KID: *Doesn't that mean you're drunk?*
> DEWEY FINN: *No. It means I was drunk yesterday.*

Or as Clement Freud, the nephew of Sigmund, put it, "'Drunk' is when you have too much to drink. 'Hangover' is when some of you is sober enough to realize how drunk the rest of you is."

But on some level, you probably already knew that—whereas you might not know what an etymological newcomer *hangover* is. At the turn of the twentieth century it didn't even exist. The state of being "drunk yesterday" was known as *crapulence*, or having the *jim-jams*, or just feeling really awful. It is one of the youngest words in the English lexicon, and yet, in the mere hundred years since it was coined, *hangover* has become ubiquitous in describing a condition that is older than language.

People have been getting drunk since the dawn of history. From the Bronze Age to the Iron Age to the Jazz Age, empires have fallen, wars have been waged, civilizations enslaved—all because of hangovers. Yet, in reading what's been written about them, the thing you'll read most often is how little has been written about them. Whether booze-soaked *Beowulf*, the liquored-up *Iliad* or a thousand drunken Arabian knights, as Barbara Holland puts it in *The Joy of Drinking*, "Nobody discussing those Herculean bouts of yore mentions the hangover, and our ancestors didn't even have a word for it."

In his massive compendium of writing about drinking, aptly titled *The Booze Book*, Ralph Schoenstein introduces Kingsley Amis's essay "On Hangovers" with two short lines: "The Literature of hangovers is small. In fact this is all I was able to find."

It is almost as if hangovers didn't exist until the fateful word to describe them. Or were so omnipresent that to write about them seemed needless, like mentioning that a character is still breathing every time she speaks a word. But it's not just the poets and historians who have, for whatever reason, ignored the hangover through history. So have the pros in white coats.

Though it's one of the most common and complex illnesses known to man, there have been practically no state-sponsored attempts to address the hangover as a legitimate medical condition—the explanation being that it is a malady for which the victims have only themselves to blame. And while that might be true—the you-did-it-to-yourself-ness of it all—one would think that even medical experts have fallen hiccupping off their moral high horses enough in the past few thousand years to try and make a go of it. But to this day, there are far more entrepreneurs than doctors digging in—extracting grape seeds, peeling guavas, mulching prickly pears, then bottling it all to line the shelves of convenience stores and surround cash registers like tiny hopeful soldiers. And where it might end is anyone's guess—much like this quest of ours.

And so to the other part of this title. While I'm the *One Man*—for better or worse—the *Quest* itself is still up for grabs. It will involve some real, fundamental research: talking to very smart people, squinting at scientific studies, compiling current data, learning about chemistry and all that, in an effort to understand what's out there. But even more so, it will depend on my "applied research"—and that's where things are sure to get dodgy.

From the lowlands of Vegas and Amsterdam to the highlands of Scotland and the Rockies; from a Canadian Polar Bear Swim to the pools of an Alpine spa; from the world's first Hangover Research Institute to an Oktoberfest Hangover Hostel; from a voodoo church in New Orleans to the London office of a doctor who has announced he has made synthetic alcohol; from those trying to research a remedy to those who say they've already found it—neither the quest nor this book will truly be finished until I've found my best concocted cure.

Sitting on my desk, next to the almost-empty bottle of (insert sponsoring brand), is a monstrous stack of little books, most of them oddly square in shape and published within the past decade, including *Hangover Cure*, *The Hangover Cure*, *Ultimate Hangover Cure*, *Cure for a Hangover*, *Cure Your Hangover*, *How to Cure a Hangover!*, *How to Stop a Headache and Cure a Hangover*, *Hangover Cures*, *Hangover Cures (Miracle Juices)*, *Natural Cures for Hangovers*, *Real Hangover Cures*,

Hangover Cures for Hungover Heads, *10 Ways to Quickly
Cure a Hangover*, *The Hangover Handbook: 15 Natural Cures*,
40 Cures for Hangovers, *50 Hangover Cures*, *50 Ways to
Cure a Hangover*, *Hangover Cures (52 Ways)*, *The Hangover
Handbook: 101 Cures for Humanity's Oldest Malady!*, *The
World's Best Hangover Cures*, and *A Little Book of Hangover
Cures*. And yet, none of them, as far as I can see, brings any-
thing new to the literature of hangovers, let alone one single,
actual *Cure*.

Treatments, maybe. Balms, soothers, pick-me-ups, hairs of
the dog, words of advice and a thousand Hail Marys, for sure—
but a real-life, bona fide cure? If that were here, I'd be on to a
second bottle by now, with a whole other book to write.

My point is this: when it comes to hangovers, both books
and people tend to use the word *cure* lightly. So I'll try to keep
in mind one particular truism; it's oft been ascribed, in all sorts
of meaningful publications, to the greatest hangover writer ever:
Sir Kingley Amis. Yet no one I've found so far—not his official
biographer, nor even his renowned novelist son—can identify the
source of this supposedly famous Amis quote:

> *Like the search for God, with which it has other things in
> common, the search for the infallible and instantaneous
> hangover cure will never be done.*

So whether he said that or not, I'm in for a helluva challenge—
but I'll give it a shot.

In fact, better make it a double.

WELCOME TO YOUR HANGOVER

You tumble from dreams of deserts and demons into semi-consciousness. Your mouth is full of sand. A voice is calling from far away, as if back in that blurry desert. It is begging you for water. You try to move, but can't.

And now that call is getting louder, like a pain in your head. A headache . . . But no, oh no, this is so much more—something terrible and growing. It is like your brain has started to swell, pressing against your cranium—eyes pushing out of their sockets. You cradle your head, in shaking hands, to keep your skull from splitting . . .

But in truth your brain isn't growing at all. It is, in fact, drastically shrinking. As you slept, your body, bereft of liquid, had to siphon water from wherever it could, including from those three pounds of complex meat that hold your messed-up mind. So now your brain, in the awful act of shrinking, of constricting, is pulling at the membranes attached to your skull, causing all this goddamn pain, tugging at the fibers of your very being.

Alcohol is a diuretic. And out with the H_2O went all those other things—electrolytes, potassium, magnesium—that make your cells (i.e., *you*) actually function. So that persistent call from your dried-out brain has a point: *You'd better get some water!*

With Sisyphean effort, you raise your head. The room begins to spin. The bar last night was spinning too, and not in a fun, disco-ball way. More like being trapped on a hellish carousel. When you closed your eyes, it just got worse—up and down, faster and faster on some devil's spinning pony.

The cause of all this whirling around (apart from the booze you drank) happens to be a fish that crawled onto land 365 million years ago and became the physiological precursor to all animal life, including ours. Its fins became talons, claws and fingers. Its scales became feathers, fur and skin. And its jawbone, containing a mysterious gel that's older than time, became your inner ear, wherein today you have microscopic hairlike cells measuring the movement of that gel, sending messages to your brain regarding sound, the tilt of your head and acceleration. And that's why the world is spinning. It is, essentially, a kind of landlocked seasickness.

Booze is like a pirate. It likes adventure—to go with the flow for a while, then suddenly take command, and also stir shit up a bit, especially once it reaches your inner ear. Alcohol is much lighter than the weird old gel in charge of your equilibrium. Unable to mix, to come to terms, the booze gives chase, around and around, until your brain thinks you're spinning out of control. When this happens, your body tries to find a fixed point—a spot on the imagined horizon. Last night, when you shut your eyes, hoping for the spinning to stop, your pupils kept darting to the right—tracking a point that wasn't there.

And now, the morning after, most of the booze has left your body; what remains is burnt out and broken down and escaping through your bloodstream. So now the chase in your inner ear is going in reverse, the world spinning in the opposite direction— your eyes twitching to the left this time. This is one of the reasons why police at roadside safety checks shine a light in your eyes. In observing the direction of your pupils, they should be able to tell if you are drunk, hungover or hopefully neither.

Not that you care about that right now; spinning is spinning, and you'd like it to stop. Sure, you might have drunk too much, but this part is hardly your fault. It wouldn't even be happening if that stupid old fish had contained a different gel—or *just stayed in the water where it belonged.* Okay, now you're getting irritable— even a bit irrational. A lot of that has to do with exhaustion and a rebounding of stimulant. You may have passed out, but not in any restful way. Once the sedation dissipated, there was no chance of reaching those deep and deeply needed levels of sleep. As much as a hangover is dehydration, it is just as much fatigue.

So even now, with the call for water like static thunder, you drop back down, thinking maybe, just maybe, you can fall asleep and dream instead of drinking in the desert. This time, though, when you close your eyes, the spinning moves downward. And now you feel your guts.

At some point last night, the booze pushed right through the lining of your stomach, inflaming the cells and making a surplus of hydrochloric acid—the same stuff used to peel paint and polish stone. So on top of the dehydration and fatigue, you've got a gut full of industrial cleaner. And your stomach cells aren't the only ones on fire. The rest of your organs are inflamed as well, swelling and tightening the tissues of your kidneys, your pancreas, your liver, and so on—impeding their ability to release toxins or absorb nutrients and water, even if you manage to get some down. To be fair, though, it's not just the alcohol that'll make this morning so rough. It's what your body's been doing to fight it.

Your liver is central command when it comes to destroying poisons in the body. To deal with your intake of alcohol, it sent out kamikaze troops called free radicals. Mission accomplished, they should have been neutralized. If, however, you kept on drinking, the free radicals just kept on mobilizing. So you might have won the battle, but now you've got rogue killers roaming through your body, looking for fights wherever they can . . .

In a desperate attempt to rein in the radicals, to regain control, your liver is kind of freaking out—and the result is a buildup of acetaldehyde. This is the same way that one of the meanest drugs ever created works. Antabuse was developed to treat severe alcoholism. When mixed with booze, it causes headaches and vomiting so extreme that even the most die-hard drinker becomes terrified of another sip. For decades, the only medical treatment for alcoholism was a prescription for instant, crippling hangover—a little taste of which you've got right now: pain and nausea until your brain stops thinking of water and begs for mercy instead.

But of course, that is all just physical; the worst is yet to come. Attempting to go fetal, you roll onto something. It feels like a fish, but it is your soul. And your squishy soul is moaning and laughing, as though you did this to yourself. Which, of course, you did.

There is rarely a time that people knowingly make themselves so quickly ill as when they get drunk or high. That's part of why, as the physical effects change, the metaphysical trauma will spread. Just as the quality and quantity of the spirits consumed may dictate the physical aspects of your hangover, the spirit in which you consumed the spirits will often decide the metaphysical. It's what makes an "I won the Oscar!/Super Bowl!/lottery!"–induced hangover and an "I lost my job/girlfriend/a thousand bucks at the blackjack table" hangover feel so very different. The one you have now is the latter kind. And eventually the pain and nausea will be a welcome relief from the thoughts swirling around in your head like antediluvian gel, or goddamn desert demons:

You've squandered your potential.
And another day of your life.
You'll never find another girlfriend.
You probably have liver cancer.
And will end up dying alone.
But right friggin' now, you just need to throw up.

Welcome to your hangover.

PART ONE

WHAT HAPPENS IN VEGAS

IN WHICH OUR MAN ON THE GROUND DRINKS A LOT,
DRIVES A RACE CAR, SHOOTS A SAWED-OFF AK-47 AND
GOES TO HANGOVER HEAVEN. APPEARANCES BY NOAH,
DIONYSUS AND AN EIGHT-POUND HAMBURGER.

*"Oh, God, that men should put an enemy in their
mouths to steal away their brains."*
—WILLIAM SHAKESPEARE

THE GLASS IS LARGE AND TWISTY. THE OLIVES ARE massive and stuffed with cheap, yet pungent Stilton that oozes down the plastic sword, creating a drifting layer like sea foam. But more baffling than this drink is what the hell I'm even doing here: trying to get drunk in time to get sober in time to do things you would *never* want to do with a hangover. I take a big sip.

It's not just the obvious ingredients—bachelor parties, free booze and a blockbuster movie trilogy—that make Las Vegas the undisputed hangover capital of the world, but a far more complex cocktail of geography, biology, meteorology, psychology, pop-culture philosophy and alcohol bylaws.

From the flight attendants' carefully cheeky landing jokes to the omnipresent "What happens here, stays here" motto to the celebration of its gangster creation myth, you are handed a line upon arrival as light and bright as a lei around your neck: "The normal rules do not apply!" So the usually sedate conference-goer grabs one of those giant fluorescent test tubes of booze on her way to check-in at 10:15 in the morning. Then the day is a string of complimentary drinks through endless rooms of flashing lights, manufactured oxygen and cigarette smoke. It is Vegas, after all. The normal rules do not apply . . . though no one told your liver.

I've experienced the Vegas effect a number of times, but still can't let go of this nagging worry: that now, when it really counts, I might not get hungover. Which brings us to this bar, and this complicated martini. I take another sip and try to focus.

The idea was to combine two assignments. This is something freelance writers often do. It helps, of course, if the two stories are somewhat compatible: a quick piece on digestifs for *Digest Digest* while on a wine tour for AeroFrance's in-flight publication, for example. But what *I've* decided to do is combine g-force with hangovers.

I am in Las Vegas for this book, but also on assignment for a men's magazine. For the book, I am researching a place called Hangover Heaven, which will involve me getting drunk enough, again and again, to put "the world's foremost hangover doctor" to the test. For the men's magazine, I will be piloting a fighter plane in a mock dogfight at six thousand feet, jumping off a thousand-foot building, zip-lining down a mountain, shooting machine guns and driving a race car—all part of an "Extreme Vegas" publicity junket. What could possibly go wrong?

As it is, I have just twelve hours to get drunk, hungover, then straight again before negotiating a ten-turn track at 150 miles an hour. My math is not nearly good enough to know whether this is even possible, but I figure three ounces of vodka and two cheese-stuffed olives are a good place to start. I study my dwindling martini and try to assess whether it was shaken or stirred. A study published by the *British Medical Journal* concluded that shaking a martini is more effective in activating antioxidants and deactivating hydrogen peroxide than stirring one—supposedly lessening a double-O agent's chances of getting cataracts, cardiovascular disease and hangovers.

There is a clanging sound behind me, bells and whistles, then shouting as someone hits a jackpot. "Another?" says the waitress.

"Yes," I say, but ask her to hold the cheese.

To be honest, I'm already a bit hungover. Due to an early flight from Toronto, I haven't slept off last night's drinks, and my gut has been bad since flying over Nebraska. I'm meeting some other journalists and our Vegas host for dinner in half an hour. But I don't know how boozy a meal it will be, and I am hesitant to tell them about my ulterior motives. At some point, I'll probably have to; our schedule is so full of dangerous stunts that, without their cooperation, I don't know how I'll manage each day to get as drunk, then sober, as required. I already feel drained, and as if half my stomach is back in Canada. There's a lot riding on this next martini.

A girl comes by wearing a pillbox hat, with a tray full of goodies around her neck. I buy a pack of Camels, a roll of Rolaids and a lighter. I chew the Rolaids, and the martini is here by the time I've lit my smoke. I take a sip.

This one is excellent: a little smoky, a little dirty, bare-knuckled and cold. And suddenly, my gut doesn't feel so bad. The oxygen they're pumping into the casino—to keep people gambling and drinking and gambling—is finally reaching my lungs. I kick up my feet, soak it in and order one more, just to be safe.

It's good to be back.

"EXTREME VEGAS," it turns out, refers not only to the driving, flying and falling, but also the eating and drinking. As such, the giant scallop and raw beef appetizers come with a tasting flight of single-malt Scotches. And they're mighty big tastes.

The main course includes five kinds of wild game. When I ask for a glass of full-bodied red, I'm brought a bottle instead. While drinking it, I explain to my dinner companions how convenient this all is—that I'm actually getting drunk! I start to tell them about my book . . . but suddenly, one of the other journalists, a travel writer from New York, just wants to talk about accident insurance—and whether we're all going to be operating motor vehicles on the same track at the same time in the morning.

I swear he's looking at me when he asks this, and he agrees that he is. Our host suggests we decide on dessert. It's almost midnight and we're scheduled at the racetrack for 9 a.m. I look at my watch, attempt the calculations and order a Grand Marnier.

It is perhaps worth mentioning that two of the subjects I've tackled most actively during my writing "career" are drinking and gambling, which might suggest that those are fairly central preoccupations—which, in certain circles, at certain times, can certainly be seen as a problem. I'm not saying I'm a problem gambler, or even a problematic drinker. It's just worth mentioning, is all . . . especially on our way out of the restaurant and toward the poker tables, where the drinks are free.

Because here's the thing: when you're writing a book about hangovers and paying for your own research, and free booze is available (but only on the condition that you gamble while imbibing it), well, then, isn't it financially and professionally irresponsible *not* to gamble, at least a little? To answer this mostly rhetorical question, I do a brief cost-benefit analysis, which I then cross-reference with a very general understanding of probability.

And this is what I find: the chance of losing money at the poker table is somewhere in the high double digits, whereas the chance of getting more drunk by drinking alcohol at the poker table (which contributes directly to my professional endeavor and therefore my eventual livelihood) is a solid 100 percent. Clearly, I have no choice but to sit down.

It is no-limit Texas hold'em, blinds ten-ten. A waitress comes by and I ask for a whisky and a beer. She tells me they can only bring one drink at a time. I ask her for a whisky and *then* a beer, and pre-tip her a ten-dollar chip. The dealer pats the table.

I post my blind, then sit back and wait for the cards, trying to gauge my drunkenness, but I'm just not sure; part of me feels disturbingly sober. Other than that, everything's going just fine—the waitress circling back, the cards landing on felt . . .

THE MORNING AFTER THE DAWN OF TIME

Hangovers predate humanity. That's at least as safe a bet as evolution—or, if you prefer, the Garden of Eden. Just leave an apple in the right place for long enough and see what happens to the birds and the bees, let alone the snakes and the apes. As long as flora and fauna have existed, so has fermentation.

If we're to go with evolution, surely our prehistoric ancestors staggered around drunk long before they could walk upright. Such debaucheries might have been rare, festive, sometimes terrifying accidents—but it is fair to assume that the world's first hangover came shortly after the world's first drunkenness. Since alcohol predates the written word by thousands of years, however, any records are relegated to the realm of ancient storytelling. And in most origin myths, it was the gods, not the beasts who first suffered the woes of fermentation—and in so doing changed the start of human history.

In African Yoruba mythology, the god Obatala got bored one day and started making humans out of clay. Then he got thirsty and started drinking palm wine. Then he got drunk and made such a mess of things—molding a bunch of his brand new humans with deformities and such—that the next morning, in typical hangover parlance, he swore off booze forever (a very long time when you're a god).

The Sumerian water god Enki was a perfectly imperfect embodiment of the dichotomy of alcohol. Somewhat fishy in appearance, he was a walking, swimming contradiction—the god of wisdom and knowledge, but also a careless, drunken letch. As such, he tried to take advantage of Inanna, the goddess of sex and fertility, by getting her plastered. But Inanna drank Enki under the table and tricked him into handing over his *me*—the governing laws by which he planned to subjugate mortal life. The next morning, realizing his drunken stupidity, he chased after her, running down the riverside while puking on his own feet. But it was far too late; humankind had acquired free will, and Enki a god-sized hangover—complete with immortal regret.

Early Israelite tradition, as well as later Christian and Jewish belief, has it that the tree of knowledge was actually a sacred vine and the forbidden fruit was a bunch of luscious grapes. When Adam consumed them, he felt himself become enlightened, powerful. And it was this godlike high that caused his fall from grace—to this lowly, fallible plane: mortality born of the first human hangover.

In the Hebrew tradition, just as Adam was being cast from heaven, he quickly cut a piece of the vine that caused his downfall. And it was this very cutting that Noah then cultivated upon the Earth—a divine gift, but given and received without the immediate knowledge of an apparently semi-omniscient God.

Most scientists and creationists agree that about ten thousand years ago there was a great flood across the Earth—and also that viniculture, the making of wine, was invented shortly thereafter. In several ancient texts, the end of the Great Flood corresponds directly with the advent of profane drunkenness. Whether it be Kezer of prehistoric Siberia, Deucalion (whose name literally means "sweet wine") of Greek mythology, Utnapishtim in "The Epic of Gilgamesh" or Noah of the Old Testament, the first thing these survivors did after finally parking their arks was learn how to make booze.

And then things got complicated. According to the Bible, Noah got so drunk on his first batch of wine that he passed out, sprawled and naked. And then, waking to discover that his son Ham had found him like this, he flew into a rage and punished him. Or rather, he punished one of Ham's four sons, Canaan—condemning him and all his descendants to eternal slavery.

Surely, there once was more to this story. Biblical scholars have long kibitzed about what really went down on that first drunken night and morning after—and why Noah, who started it all by getting so drunk, suffered none of the wrath of God. Noah's defenders explain it like this: as the first human to ever get drunk, he couldn't rightly be judged; after all, how do you avoid debauchery if you don't even know it exists? Imagine being the first person to ever get drunk, and then hungover.

The great Kingsley Amis states, as his primary hangover imperative: "Start telling yourself that what you have is a hangover. You are not sickening for anything, you have not suffered a minor brain lesion, you are not all that bad at your job, your family and friends are not leagued in a conspiracy of barely maintained silence about what a shit you are, you have not come at last to see life as it really is."

But what if you didn't know that? What if you didn't know anything about any of this? You'd think you were poisoned and dying, going mad and to the devil. Even someone less stressed-out than Noah might have lost it just a little. And also, who's to say that God didn't hold him accountable after all, just as he did Adam? Perhaps the original biblical drunks *were* punished, in that Old Testament style, not just with the first-ever hangover, but hangovers for all mankind—forever and ever, amen . . .

WHAT HAPPENS IN VEGAS (THE MORNING AFTER)

I gasp awake to a buzzing kind of ringing and a blaring kind of beeping. I grab the phone, slap the bedside radio and fall back into position, curled yet sprawled.

And then slowly I begin to remember: *plane to Vegas, to martini, to dinner, to poker, to . . . getting a little fuzzy.* I creak my eyes open and look around. There's something odd . . . My thoughts are leaping forward. *Drive a sports car, see the hangover doctor, get dressed—wait: other way round.*

I'm trying to focus my eyes, but there's still something . . . something *off* about this room: there seems to be more space. And everything's in a different place. I get out of bed, limping a bit, and locate the button for the blackout blinds. As they rise, I stagger back, now looking down—much higher than I expected—on a giant roller coaster. It has appeared out of nowhere. Faces loop-the-loop below me.

This. Isn't. My. Room.

I turn quickly—but I'm alone. And my stuff all seems to be here (though not where I'd have put it). So maybe this *is* my room—just twice as big as I remember, and higher up, and on the other side of the building . . . My mouth is very dry.

I get a glass of water from the enormous bathroom, glug it down, fill it again, then sip more slowly . . . and now things are coming back in flashes. Words, images and people hyphenating like narrative synesthesia: *forever-lamp, telephone-anger, pink tie-douchebag, barefoot-security, cleaner-Rosalinda* . . .

Blackouts, of course, can be a sign of dangerous drinking or a neurological problem—and, for some, a terrifying reality: a gaping unknown into which the hangover tumbles. I, however, tend to remember everything, even if it takes a bit of time—which, at the moment, I do not have.

The alarm starts a post-snooze ring and I locate my pants. As I pull them on, I feel a sharp pain in my right foot, but there's no time for that—nor this mysterious hotel room upgrade. Judging from the contents of my pockets (a folding corkscrew, a small doorknob, a stack of scribbled notes, several ATM withdrawal slips and zero cash), it probably wasn't in celebration of a winning streak.

I limp out of the hotel and into a waiting cab. "Where to?" says the driver.

"Spearmint Rhino."

He chuckles. "Keep the party going?"

Spearmint Rhino, located on the far edge of town, is one of Las Vegas's most notorious strip clubs—open twenty-four hours a day, 365 days a year, so that any day can start with bottle service and a simple lap dance. Across the street is Hangover Heaven.

OPEN LESS THAN a year, Hangover Heaven advertises itself as "The only medical practice in the world dedicated to the study, prevention and cure of hangovers."

Its treatment packages include Sunday School ($45), the Redemption ($99), the Salvation ($159) and the Rapture ($199). Their website is full of testimonials on the importance of treating the hangover as a legitimate medical condition. Then, clicking on the merchandise page, you can purchase ball caps, shot glasses and T-shirts that read, I FEEL LIKE JESUS ON EASTER MORNING.

Laid out before me is one of those single-floor, flat-roofed business/industrial parks that, at least on TV, tend to house bounty hunters and self-improvement gurus. I'm starting to imagine the business of treating hangovers as a congenial mix of the two. I find the door and open it.

"How you doin' today?" says the smiling girl leaning against a trade-show style counter/desk thing.

"Fine," I say, then quickly remember the purpose of my visit. I don't want her to think I haven't taken this seriously: "I mean, considering the, you know . . ." I lift my hand and give the universal hang-five sign for drinking booze. Despite the vagueness of my memory and a general feeling of crapulence, I'm still not sure I drank enough to really test this place.

"I'm Sandy." She picks up a clipboard. "You the writer guy?"

"Yes, ma'am," I say. And now the effort to come off as adequately professional and hungover at the same time is starting to hurt my head, which I take as a good sign.

"Well, don't you worry," says Sandy, in the middle of the desert. "We'll get you right as rain."

Hangover Heaven is the swollen brainchild of Dr. Jason Burke, who claims to have "treated more hangovers than any other doctor in the world" by applying his experience as a post-operation anesthesiologist to post-partying problems. I've talked to him on the phone. The North Carolina accent fit his photo so perfectly I found myself squinting at the imagined gleam off his unearthly white teeth. But I am going to have to wait a little longer for a face-to-face meeting. By the time he gets into work today, I should be speeding around a racetrack.

As it happens, the guy who will be treating me in Dr. Burke's stead works part time as an EMT at that very same racetrack: "I just sit around, waiting for them to crash," he says. "Oh, I don't mean *you*. And don't worry—they never crash." This all seems highly suspect, and it's making me feel unwell.

"So, how do you feel?" says EMT Paul as he adjusts my giant leather La-Z-Boy, one of six in this white-walled room.

"Unwell."

"How bad overall, on a scale of one to ten?"

"Seven and a half," I say. But I'm worried I might be skewing it a bit high, out of guilt that I'm not hungover enough.

"What do you think all that g-force'll do to your hangover?" says Paul, stringing up an IV bag.

"I don't really know," I say. "But I'm not going to *have* a hangover, right?"

"That's right," he says, and taps the inside of my arm. "Make a fist."

According to Hangover Heaven's in-house data, their success rate is 98 percent. What's in this IV is called a Myers' cocktail. It includes electrolytes, magnesium, calcium, phosphate, vitamin C and a bunch of B vitamins, and it's meant to promote hydration and alcohol absorption. Paul adds in Zofran for nausea, an anti-inflammatory called Toradol and a steroid called dexamethasone. Then he gives me a "Super B" shot in the shoulder. This, he says, will help prevent hangovers in the coming days. I'll also be given two pill bottles of supplements, one to be taken with lunch, the other with dinner.

While Paul's explaining all this, his colleague Greg, a registered nurse, is preparing the oxygen tank. They are both big, bullet-headed guys, and very cordial. Like Sandy, they could just as easily be working in a nightclub. That, of course, is the Vegas way: a welcoming manner, an easy laugh and lots and lots of oxygen.

Greg puts the mask on me. "We'll leave it there for half an hour."

"How about a movie?" asks Paul.

"That's okay," I say, sounding like an astronaut inside this mask. They look disappointed, Paul already halfway to the big-screen with a DVD in his hand. I can see the case, and realize this is supposed to be a fun part of the experience—watching *The Hangover* while mine is being cured. I don't want to disappoint, so I make a gesture like I've changed my mind.

As it happens, the first time I watched *The Hangover* was also during a trip to Vegas, and also while being treated for a hangover. But that was a *real* one—the kind where you almost die, lizards drink from your eyes and your doctor girlfriend has to put you in a cold bath to bring your temperature down. It's the kind of story I'll save for a chapter on worst hangovers ever.

"It's not working!" says Paul, scrubbing the DVD on his scrubs and trying again.

"Isn't there another copy?" says Greg.

"It's cracked."

"Dammit."

I assure them it's alright; a good movie, but I've seen it before. They put on *Ted* instead, about a pothead teddy bear. On the wall beside the TV is a large poster with the Hangover Heaven bus floating through the clouds. PRAYERS ANSWERED!, it reads. The lights go out, and they leave me to the cure . . .

AN HOUR OR so later, I join the other journalists in the briefing room of Vegas Dream Racing—where, for enough money, you can drive one of the fastest cars in the world. Despite the morning treatment, I am feeling somewhat shaky.

"You ever done this before?" asks the writer from Iowa—an old-school, avuncular journalist who's self-published a finely crafted, nicely jokey self-help book available through his website on a pay-what-you-can basis. It is called *Use All the Crayons* and has sold more copies than everything the rest of us have authored combined. He is as self-deprecating and inspiring as the New York travel writer is not.

"Nope," I say, "but I always wanted to." I tell him about when I lived in a little town in Italy, where every Ferrari is made. "You could hear them racing on the test track all day—that thundering, growling, zooming Doppler thing. It was awesome."

"Cool!" he says. "I'm kind of freaked out. But in a pretty good way."

As we head over to the simulators, it occurs to me how close I could have come to blowing this opportunity. There are, of course, myriad factors that contribute to the intensity of a hangover. I've been messing with altitude, time zones, climate, drinks with cheese in them, whisky flights, five kinds of wild game, poker, cigarettes and a severe lack of sleep. Even *I* realize that's a lot of faith to put in a medical practitioner who also sells shot glasses.

The simulator is tricky, and there's still that pain like a splinter in my foot. When I spin off the track, the seat vibrates. I hit a wall and the whole thing thuds. The instructor keeps assuring us that the real thing is easier. But the real thing is also *real*: an earthbound rocket with a single steering wheel and a solitary brake pedal—and one false move . . .

After signing all the waivers and donning my race-car-driver jumpsuit, I start to feel a little queasy. I find the washroom, look at my eyes in the mirror, then splash cold water on my face. There is a pounding in my temples, but it feels, somehow, far enough away. I think of all the things I've done with a hangover. *This,* I tell myself, *is nothing. Tomorrow is fighter planes, then jumping off the Stratosphere. And anyway, you went and got the cure, remember?*

I nod, grab my helmet and go to find my race car.

THE CONTORTIONS I go through to squeeze into the tight cage of the driver's seat make it feel like getting behind a wheel for the very first time. As worried as I am about crashing, I'm more concerned about not having the guts to go as fast as I possibly can. But then I turn on the engine, and all of my worries just dissipate—along with any hangover symptoms. They'll return tenfold later, in the limo back to Vegas. But right now, all I feel is this

motor. The rumbling is deep and powerful and just barely contained—like having a dragon on a leash, but also being inside it.

I pull out, take a corner and then another, my brain trying to catch up with my body, which is trying to keep up with
the car, only part of me listening to the instructor, who is sitting in the passenger seat. I know what I'm supposed to do: not
touch the brakes, downshift into the curve, then gas and upshift,
upshift . . . but I'm still trying to get the feel. Then we hit the first
straightaway.

I push it up and up, and the rush is incredible. And now it's
the second lap, same first turn—my non-hangover is gone, and
suddenly I get it: this *is* easier than the simulator. In fact, it's
easier than a normal car. You turn that wheel and it does exactly
what you want—no compensating, no worry of fishtailing, no
braking before or correcting after. You just trust it and do it—like
using the Force, but with your eyes wide open.

That's how I take the next turn, both holding fast and letting
go. And when I hit the gas, there it is: that thundering growl. It
fills all five senses and moves straight through me like an adrenaline riptide.

I reach 160 miles per hour. The g-force is like being hugged by
a ghost, both in and out of time. It's the best I've felt in months.

"A HUGGING GHOST?" says the young California freelancer as
we head back into Vegas for lunch. "That's kind of weird." The
racing is long over, and now we're in one of those giant party
limos, some sort of mariachi club music booming out of the
speaker, and I'm in a seat facing backward.

"A really strong ghost," I say, by way of clarification. But I
am, in fact, starting to feel kind of weird. Then, with a sort of
flip of my guts, I realize this is the worst place for me to possibly be—backward in a moving, thumping partymobile. But it

is already too late. And now the tinted windows are crushing in, my innards pushing to get out. I wrap my arms around my torso, close my eyes and try desperately to rock with—not against—the spinning, bludgeoning music.

Finally, we pull up in front of a hotel and I burst out of the limo. Iowa Crayon is calling out the name of the bar where we're supposed to have lunch, and I try to wave an "all okay." But all is not okay. All is seriously messed up. If I could produce thoughts, I'd think maybe the g-force counteracted the hangover treatment—or that no matter what you put in an IV, and how much oxygen you pump into your lungs, it's still not a good idea to move from drunkenness to a race car in fewer than eight hours. But I cannot think. Instead, all I've got is that wounded badger instinct where you're limping through the heat with nothing in mind but finding the porch—the one to crawl under and die.

I locate a washroom instead and open the pill bottle marked DAYTIME, the Hangover Heaven bus driving through angelic beams of light on the label. Inside is a chewy cube wrapped in foil and some capsules filled with God knows what. Later, I will read the label—"Taurine, 1,000 mg; Milk Thistle, 330 mg; Resveratrol, 500 mg; Acai; N-acetylcysteine, 600 mg"—but still it will mean nothing to me.

I choke them down and slurp at the water in the sink. Then I sit in one of the stalls for a while, waiting for the spinning to stop. If I *am* going to throw up, now would be the dumbest time of all—right after swallowing all those desperate, hopeful pills.

Eyes half-closed, legs half-steady, I find our lunchtime bar and join the others at the table. New York Travel glances over reproachfully. Suddenly, I have the strange sensation that I have shrunk to half my size. I close my eyes, then open them again—but it is still sitting on the table in front of me: a burger twice as big as my head.

"Extreme Hamburger!" says our host.

"Eight pounds," says the California Kid. "If you eat it in an hour, it's free!"

I want to point out that *everything* is free—we're idiot journalists on a kill-yourself-one-way-or-another junket. But that might sound ungrateful. And also, instead of words, I fear that vomit would come out.

The waitress begins to describe the various strategies by which people have eaten the hamburger and beaten the record: pickles first, licking off the mustard, dipping the bun in beer, masticating the meat, then using a spoon . . . on and on until there's nothing to do but run. I run through the bar, out a sliding door, and collapse against a golden statue of a pirate. From this angle, it appears to be Captain Morgan, who was actually a privateer rather than a pirate, and who drank himself to death in 1688. I wish, as I start to pass out at his feet, that I wasn't aware of things like this.

DIONYSUS AND THE DOUBLE DOOR

Another contender for the earliest human hangover dates back to the early days of the Greek gods, when Dionysus first began freewheelin' across this mortal plane. Yes, Dionysus was the god of wine. But first, and perhaps more importantly, he was the demigod of dichotomy. The son of Zeus and a bold, beautiful earthbound woman, he had the powers of the gods along with the unruliness of the mortals—and the pride and lust of both. The result was a creative and charismatic, amorphous and adventurous, dangerous and dichotomous force, the like of which the world had never known. It is what made him the god of wine—and he drank it like mortals breathe air.

On one particular day, he found himself sipping from a bottomless glowing sack, walking down a dusty road in the region of Pandion, singing songs as he made them up—about oceans in the sky, flying mermaids and the joy of being alone. He'd grown tired of drinking with gods on Olympus, and as he walked and sipped, he let himself change—from a demigod to a Brahma bull, to a lightning bolt, to a lizard king—and then to a lithe, smiling person as he shimmered over some farmland and came upon a home.

It was that of Icarius and his daughter, Erigone—and Dionysus took to them right away. He offered his winningest smile and some of what he was drinking, and they invited him in for a meal.

Out of respect for his mortal hosts, and knowing that only gods could handle wine straight up, he mixed their cups with water. And then they got down to having a good time—so much so, in fact, that before hitting the road again, Dionysus gave them a batch of his hooch, and the secret of how to make it. Some time after that, Icarius, Erigone and their dog, Maera, set out to share this heavenly discovery with a few of their earthbound neighbors. And that is when things went sideways.

As with any ancient story, there are many versions and interpretations, but the gist is this: Icarius's neighbors drank the wine straight up, passed out, then felt so wretched upon waking that they thought they'd been poisoned. So they beat Icarius with clubs, chopped him up and threw him down a well. The distraught dog jumped in after, and Erigone, seeing all this, hanged herself from a tree. Then Dionysus showed up.

Certainly, it's unwise to anger any of Zeus's offspring—but especially this one. By the time Dionysus was done, humanity's first drunken miscommunication had become a morning-after apocalypse. Any mortals in the vicinity had been pulverized, plagued or banished to a hellish island. And as for the murderers of Icarius themselves, Dionysus reserved one decidedly unconventional form of torment: he seduced them so thoroughly, yet incompletely, that they were driven out of their minds and made to burn forever in a state of unfulfilled sexual desire.

A much more pleasant immortality was bestowed on Dionysus's deceased drinking companions, as he turned them into celestial beings—even the pup Maera, who became the bright and lonesome Dog Star.

Soon thereafter, the king of Athens decreed that only a god could handle unmixed wine, and any mortal who tried would go mad and/or die. And as a taste for wine spread through early civilization, mixing it with water became a decisive tenet separating civil society from barbarism, wisdom from carelessness and health from debauchery.

Young Greeks were taught to drink responsibly at *symposiums*, a riff on the word *gymnasium*—suggesting that the same kind of strengthening, practice and discipline were needed to become a good drinker as to become a skilled athlete. Sipping from bowls of watered-down wine while musing on the beauty of logic and the logic of beauty required careful training by a master *symposiarch*, such as Plato. According to Tom Standage's very cool book *A History of the World in 6 Glasses*, scholars at the time observed that "those who dined with Plato felt perfectly fine the next day."

As early civilization reaped the rewards and wrestled the repercussions of wine drinking, Dionysus become so powerful—so varied in talents and gifts, and encompassing so many paradoxes—that not a hundred names could hold him. He became known as the Wild One, the Two-Faced God, the Dancer, the Loosener, the Bringer of Light, the Reveler, the Ecstatic One, He Who Makes Women Mad, the Warrior, the Emancipator, the Giver of Grace, the Holy One, the Redeemer, the Farthest, the End, He of the Double Door . . .

Like that of the coming Christ, Dionysus's mass appeal was derived from the idea that his devotees could, through acts of communion, attain salvation in the afterlife. They did this by drinking the wine that was his blood. Enough of it, and they would feel the precursor to salvation—a freeing of the soul from its earthly body. But a bit too much, and the opposite could happen: chaos, earthly degradation and a bad bout of soul-sickness. Thus the double door, leading one way, then the other, to both heaven and hell, the divine duality of intoxication and hangover.

WHAT HAPPENS IN VEGAS (WHEN SHOTS ARE FIRED)

My headache hits maximum intensity just as we arrive at the gun shop. I feel like I'm going to die, and this isn't on my bucket list. We get a basic briefing, earplugs and goggles, then an "Extreme" surprise: in addition to the standard weapons of warfare, we'll be allowed to shoot a sawed-off AK-47—which has a blast range so wide it'll actually throw flame.

My fellow journalists choose esoteric targets: a mummy in black tie, a hipster G-man, a Rambo-like clown. The New York travel writer chooses an attractive woman with a zombie approaching from behind. (It turns out he's a crack shot, and the zombie, of course, is left unscathed.) I choose the generic, nondescript target: a sort of spliced, featureless Barbapapa. And even this I have no desire to shoot.

Every bullet, every cartridge, every piece of shot feels like it's ricocheting through my cranium. God, what it must be like to go into battle with a hangover . . . pain, fear and sickness fighting inside you while death rains down all around. That soldiers have done this for millennia just makes my headache worse. I shoot until I'm nauseous again, and don't even try to aim.

After the gun range, we go to a Mexican restaurant for Massive Margaritas and Super-Spicy Tacos—my list of things you would never want to do with a hangover nearly complete in just one day. Tomorrow is aerial acrobatics in a fighter plane and jumping off a thousand-foot building. I have never looked with such loathing at a flight of free tequila. And then the mariachis descend.

AS I FINALLY stagger back into our hotel lobby, I realize I don't even know what floor my mysterious new room is on. Before I can ask, the concierge belts out a welcoming laugh. "Ah! Mr. Bishop! How are the new digs working out? I do apologize for last night!"

I don't know what he's talking about, but assure him all is forgiven and that the new accommodation is more than adequate. And could I just get the number again . . . ?

"Of course." He taps a few keys to look up the room. "And how was your day?"

"Great," I say, then limp toward VIP Tower 3.

The day's been so nauseatingly eventful that I haven't had the chance to even try to remember what happened last night. I'm thinking about it while I turn on my laptop to send a painstakingly breezy message to my editor. And there, in the middle of the desktop, is a new file labeled OKAY! I'M DRUNK! I click it open and begin to read:

> *Okay! I'm Drunk! But now the lamp won't turn off and I can't even unplug it. The cord goes right into this stupid table thing in the wall! I gotta go to sleep so I can drive the race cars tomorrow and I can't turn the light off!*
>
> *Okay, I tried unscrewing the light bulb but I burnt my hand and I broke the light bulb and now there's glass on the bed. I can't frigging believe it! I've been on hold with*

the front desk for half an hour and I don't think they even
know the call's coming from inside the hotel. The call is
coming from INSIDE *the hotel!*

Now the phone is broken!

Oh my god! The stupid pink-tie guy at the desk down-
stairs said I was only on hold for 8 minutes and he said
I was drunk just because I didn't have shoes on! What a
dick! And I had to come up here in the elevator with two
security guys and they think he's a dick too. We were all
laughing about my shoes, but then I forgot when I came
in and now I have glass in my foot and the security guys
said they'd send somebody to fix the phone and I've been
waiting for half an hour!!

I just met the nicest woman in the world. Rosalinda.
I think she's the head of the cleaning staff. She was out in
the hall. It's so much later now because I had to tell her
the whole story. She could see I wasn't even drunk. She
said she's going to get it all sorted out right now and get
me a new room and I said, make it a nice one—where I
can smoke in it, too. So now I'm waiting again. This is
the longest day. I can't believe I have to . . .

The phone is ringing. I stop reading my own drunken non-
sense and pick it up.

"How was your day?" At first I think it's the concierge again,
then recognize the confident drawl of Dr. James Burke.

"Actually, quite painful," I say.

There is a brief silence, and I realize I'm like a restaurant critic
telling the owner the fish was inedible.

"I'm sorry to hear that," says Dr. Burke. "Why don't you take
me through it, and maybe we can figure out what went wrong."

"Sure," I say, and start with the blue cheese martini . . .

"How do you feel now?" asks Dr. Burke after I've told him everything.

"Okay, I guess."

"Well, you know, we have a success rate over 90 percent—"

"The website says 98."

"Right. So I have a few theories as to what happened."

"Go for it." I open a fresh file to take notes on my computer.

"Well, for one—like I said—our success is not 100 percent. For some people, it just doesn't work. I don't know why. But that's not very likely here."

"About 2 percent likely."

"Right. So here's the thing: I'm looking at how much you said you drank, when you stopped drinking and when you were treated in our clinic. As you understand, our system has been developed to treat hangovers . . ."

"Uh-huh."

". . . And part of what constitutes a hangover is alcohol withdrawal—which I'm beginning to doubt was part of your status."

"What do you mean?"

"Well," says Dr. Burke, "it's fair to assume you were in a state of intoxication rather than withdrawal when we treated you. And for some time after . . ."

Then finally I get it: he's saying that while driving a half-million-dollar car at 160 miles an hour, I was the *opposite* of hungover. I was *hammered*. All that work to get properly pissed, and I screwed it up by applying myself too much. So now this whole experiment is a bust. And on top of that, I'm a drunk driver.

"Damn."

"We'll just have to try again," says Dr. Burke.

But now I'm thinking about the fighter planes in the morning—barrel rolls, hammer drops, nosedives . . . I close my computer.

"I think I'm just going to take it easy tonight. We'll get back on track tomorrow."

"Alright," says the hangover doctor. "I'll see you after that."

I pour myself a whisky and get undressed. The light turns off easily. When I close my eyes, the room feels tight, like I'm being hugged by a very strong, very merciful ghost.

A DRINK BEFORE THE WAR

Plato, who had learned to think and drink in careful balance from his teacher, Socrates, passed those teachings on to his student Aristotle. And then Aristotle's most notable student was the Macedonian king Alexander the Great, who followed the wisdom of ancient Greece—except when it came to drinking.

Like his father, Alexander was a passionate, drunken warrior king. And like his mother, he was an ardent follower of Dionysus—so much so that he came to see the path of his own conquests as a recreation of the wine god's travels. He and his men drank to stupor every night, riding into battle fighting headaches as much as the Hun, prevailing every time. He blew the new society out of the water, the water out of the wine—then tried to civilize the world by painting it red.

In the end, Alexander took more territory than any human before him, burning cities to the ground, sometimes by accident when the Bacchic celebrations got out of hand. Whether or not he brought civilization to the world, he certainly succeeded in spreading the hangover to its territorial limits—though not everyone appreciated the effort. As Demosthenes said of his drinking, "It is a good quality in a sponge, but not a king." And perhaps he had a point. Unbeaten in battle, it was the bottle that finally

brought down Alexander the Great, though exactly how is a matter of some debate—whether his body gave out from drinking
too much or went into shock because he stopped drinking the
way he did everything else: all at once, and very fast.

But of course, Alexander and his armies weren't the only
ones to drink and fight their way into the history books. Ever
since Homer, whose *Iliad* and *Odyssey* are full of warriors full
of wine, historians have recognized the importance of booze in
the making and sustaining of bloodshed. Norse mythology leans
heavily on the battlefield success of berserkers: juiced-up warriors in a frenzied, fearless state induced by the booze of the
gods. And not only humans went to war this way. Marco Polo,
who drank himself around the world, noted that in Zanzibar,
warriors gave buckets of rice wine to their elephants to "inflame
their courage."

During the American Civil War, when Ulysses S. Grant came
under fire for drinking too much, President Abraham Lincoln
vowed to make more booze available to any of his other generals
who had "never yet won a victory."

Of course, not all drunken warriors excelled in battle. In
Beowulf, those who've come to fight the mysterious beast Grendal
are slaughtered in the mead hall while sleeping off their cups of
courage. Barbara Holland compares the ambush to the hungover Hessians taken down by George Washington, and also the
Saxons who stayed up drinking before the Battle of Hastings and
therefore lost England to the more disciplined Normans, who, as
she puts it, were "sober, or at least less drunk."

During World War I, professional soldier and de facto journalist Frank Percy Crozier detailed battlefield horrors while
sketching almost pedestrian asides: "I saw a colonel sitting at
the entrance to a communication trench, personally issuing
unauthorised tots of rum to his men as they passed him in single

file at 3 p.m., on a fine clear Spring-like afternoon, on their way
to hold a line for the very first time. . . . Badly based on brandy,
he thought everybody else felt as he did—dejected and desolate.
Despondent and at all times *difficile* . . . the drink-drug addict
was removed to England, there to degenerate and eventually die.
The safety of the line outweighed all other considerations. Drink
control was imperative."

Although at that time the term *hangover* barely existed, there
was surely no excuse to be throwing around words like *difficile*.
But thanks to the Industrial Revolution, ideas about drunkenness
in regard to productivity and safety were starting to change—
along with alcohol's once-sacrosanct place in Britain's military.
As Prime Minister David Lloyd George declared, the Empire was
fighting "Germans, Austrians and Drink—and as far as I can see
the greatest of these foes is Drink."

The Russians, meanwhile, seemed to succeed even when completely pickled, and sometimes precisely because of it. In both
world wars, many of Germany's perfectly timed attacks were
foiled by the unpredictability of hungover Soviet troops—lost,
asleep or just late in reaching their intended destination. Despite
the accidental successes, the Kremlin eventually became so concerned by its own drinking that the KGB started working on a
pill to prevent operatives from getting too drunk. The pill didn't
work for that, but it did seem to help with the hangover . . . or
at least this is the story told by RU-21, the first mass-marketed
hangover product to come out of the Cold War, and arguably one
of the most successful in terms of American sales.

In his gripping memoir *Green Berets Gone Wild,* Russell
Mann writes about being a medic during the Vietnam War. An
important part of his job was taking care of his sergeant: "He
liked to have his medic nearby in case he got hit, but more important he wanted me around for hangover repair. . . . Outside of

combat he was a drunk, a womanizer—not a person you would like to be around; but in combat he was wonderful."

The recruitment offices have had their share of hangovers too—including those of two very different heroes: Tommy Franks and Bruce Springsteen. According to General Franks's own memoir, in 1965—after failing out of college, in a bout of depression and at the tail end of a drinking binge—he decided to enlist in the army to "jolt" himself out of "a soul-crushing hangover." Franks went on to become one of America's most decorated soldiers, eventually leading the attack on the Taliban in Afghanistan and the 2003 invasion of Iraq.

Bruce Springsteen, meanwhile, has an inverse story he's told on stage, as an introduction to the song "The River." It is about growing up in 1960s America, under the constant threat of the draft, while all that's coming back from Vietnam are coffins and broken boys. It is also about the hostility in Springsteen's boyhood home, where his father would end their fights by wishing the army upon his son—that they would take him, cut off his shaggy hair and finally make a man out of him.

When the draft notice eventually comes, Springsteen takes off with his friends. They stay up for three days until it's time to get on the bus and report to the recruitment office.

In telling this story, the bottom drops out of Springsteen's voice when he says the word *scared*. He never says the word *hungover*—only that he went, took the physical . . . and failed.

The crowd breaks into applause, even as Springsteen mutters that it's nothing to cheer about.

Then he recalls the moment of going back home, after three long days, to where his father sits waiting in the sunlit kitchen. The boy tells his dad the army wouldn't take him. There's a long, silent beat before his father finally speaks.

"That's good," he says.

Then, on the stage, it rises out of Bruce Springsteen, now grown to be a man: some of the most grieving, echoing harmonica you've ever heard. It sounds like pipes over poppy fields—a small part of an empire's soul aching for redemption, the rest of it drenched in booze and blood.

PART TWO

WHAT HAPPENS
ABOVE VEGAS

IN WHICH OUR MAN IN THE AIR FLIES A FIGHTER PLANE,
JUMPS OFF A BUILDING AND KEEPS ON DRINKING.
APPEARANCES BY CHUCK YEAGER, CAPTAIN HADDOCK
AND DR. JASON BURKE.

"Half a hangover merely wrecks the night before
as well as the morning after."
—CLEMENT FREUD

I AM SWEATING IN AN AIRPLANE HANGAR IN THE Nevada desert, among biplanes, pool tables and a fully stocked bar. The jukebox offers Kenny Loggins as well as the Everly Brothers—all part of Sky Combat Ace's Top Gun Experience.

Pilot Richard "Tex" Cole is describing what we're about to do. He is talking hammer drops and evasive maneuvers and how best to get behind your opponent: "The fastest roller coasters in the world can put 3.5 g-force on you. But the baby you'll be riding—she's a ten!"

I am not in the mood for any of this: not the bravado briefing and fake macho bullshit, not dogfights real or mock, not aerial acrobatics, not g-force and barrel rolls six thousand feet above a dry lakebed, and definitely not hammer drops. I don't even like roller coasters.

One of the team-building things we're supposed to do is give each other call signs—like Iceman or Goose—by choosing a nametag from a board on the wall. The California freelancer grabs "Toxic" and hands it to me gleefully.

Waking this morning felt worse than it did yesterday—which, according to Kingsley Amis, is a very good sign. It is, in fact, the first of his eleven steps for dealing with "The Physical Hangover": "Immediately on waking, start telling yourself how lucky you are to be feeling so bloody awful. This recognizes the truth that if you do not feel bloody awful after a hefty night, then you are still drunk and must sober up in a waking state before hangover dawns."

So, apparently, I'm lucky. But I'd have been luckier to think of this yesterday—before the IVs and the race car, the limo and the lunch, the margaritas and machine guns. And now I've got that second-day bottle-ache. It's like when you make the mistake of trying to get back in shape: the morning after the gym might not be so bad, but then comes the day after that, and it hurts just to get out of bed, let alone fly a fighter plane.

Checking details on the website this morning, I happened to notice this: "The most common cause of airsickness at Sky Combat Ace is due to hangovers. We know Las Vegas is a crazy place, but try to hit the sack early the day prior to your Sky Combat Ace adventure!"

Not that going to bed early ever warrants an exclamation point, but it's sound advice nonetheless—and exactly what I've done. I turned in early, just like they suggested; and then, as recommended by Master Kingsley, I woke up feeling awful. But now this hangar in the desert is hot as hell, my mouth is dry, and it seems like maybe I've missed a step in dealing with the physical hangover.

Immediately following his eleven imperatives, Amis suggests two purported hangover cures he never got to try, as they were "hard to come by." The first involves going down a mine shaft, which sounds like a terrible idea. And the second one is this: "Go up for half an hour in an open aeroplane (needless to say, with a non-hungover person at the controls)."

According to Sky Combat's brochure, they happen to offer precisely that: a forty-five-minute flight over the Hoover Dam in a classic open-cockpit biplane, with the guy in the back doing all the piloting. This sounds like a much better plan—and a perfect opportunity to test one of Kingsley Amis's untested cures.

Except that Tex says, "Nope. No can do." The winds are too high for the Hoover Dam Experience. It's better we stick with aerial acrobatics while trying to shoot each other out of the goddamn sky; apparently, the winds are just *perfect* for that.

Before suiting up, I visit the gift shop, where they happen to sell Dramamine. I know it can make you drowsy, but in the middle of a dogfight I'd rather be yawning than vomiting. I walk out onto the tarmac, helmet tucked under my arm, aviators glinting in the sunlight, just as New York Travel is getting out of his plane. I ask him how it was. He gives me the thumbs-up but doesn't look me in the eye. At the Cirque du Soleil show after dinner last night, he asked the little boy in front of us to be quiet. A little boy at a circus. I wish I was dogfighting him instead of this friendly freelance kid.

I climb into the cockpit. Hollywood, the instructor, is sitting behind me, and his voice comes through on my headset. Fittingly, he is giving me more tips on how to look good on screen than how to fly. Apparently, there is a cockpit camera that will catch my every yelp and grimace. Burbank the freelancer (from California, but not from Burbank) is now rolling down the runway in front of us.

"Better to take your shades off," says Hollywood, "so we can see the whites of your eyes." I doubt they're very white. Surely the Dramamine has kicked in by now, but instead of better, I'm starting to feel worse—sweatier, shakier, edgier—and now the propeller is starting to spin . . .

Not until months later, laid up in hospital for nothing to do with fighter planes, will I learn an important fact: I have an allergy to Dramamine, which happens to cause many of the symptoms of an extreme hangover—nausea, sweating, anxiety, muscle pain—as well as irrationality, heart palpitations and hallucinations.

"Ready for takeoff?" says the voice in my head. "Over."

I give a thumbs-up, just like Maverick would.

IN HIS BOOK *The Bonfire of the Vanities*, Tom Wolfe describes the "membranous sac" of a journalist's head that contains "the yolk, the mercury, the poisoned mass" of his brains. Should he try to get up, it would "shift and roll and rupture the sac." It is one of the most lauded hangovers in modern literature.

But *The Right Stuff*, Wolfe's chronicle of US fighter pilots (and eventually astronauts), has just as high a blood-alcohol content. Wolfe describes your everyday ace on an average morning "waking up at 5:30 a.m. and having a few cups of coffee, a few cigarettes, and then carting his poor, quivering liver out to the field for another day of flying."

Apparently, when he cracked the sound barrier for the first time, the great Chuck Yeager was feeling a lot like I am now. According to Wolfe, he'd fallen drunk off a horse a couple of days earlier, and so came to that historic flight with broken ribs on a second-day hangover. Unable to do so otherwise, he used a sawed-off broom handle to pull the cockpit door closed. And the rest is sonic-boom history.

WE'RE FINALLY AIRBORNE. Hollywood passes the controls to me. I'm feeling a little better now. Flying, after all, is intrinsically liberating, and there is so little to hit up here. I could be a pilot. Hell, I *am* a pilot—a goddamn sky pilot, and the sky is blue and clear. But then, I think, the sky shouldn't be *this* clear. And I remember Tex's words in the briefing room: "Lose sight and you lose the fight."

Where the hell is Burbank?

"Behind you," says Hollywood, reading my thoughts. Then he says, "Over," which kind of bugs me. I should be moving evasively, but for some reason I've also chosen this moment to take off my shades—so the camera can get the whites of my eyes—and I'm trying to hook them onto my flight suit as Burbank closes in.

"What the hell are you doing? Over."

I try to straighten up and fly crooked. And now, finally, I start doing what they taught me: spinning and rolling, diving and dropping. The fight is on, and this part isn't fun. Through hammer drops, barrel rolls and loop-the-loops, what I concentrate on most is not throwing up.

You might think that aerial dogfighting is top of the list of what not to do with a hangover, and you'd be right. But on the other hand, it is one of the few jobs that comes with a bonafide hangover treatment. As Wolfe explains, "There were those who arrived not merely hungover but still drunk, slapping oxygen tank cones over their faces and trying to burn the alcohol out of their systems, and then going up, remarking later, 'I don't advise it, you understand, but it can be done.' (Provided you have the right stuff, you miserable pudknocker.)"

I, however, have the wrong stuff: no oxygen tank, a wussy gut and a hangover that appears to be hurtling back to life. And now it's hard to know what's got me most: the g-force, the leftover booze or (in retrospect) that goddamn Dramamine. It is like I'm sweating backward—into my brain instead of out of my skull. It clouds the vision behind my eyes. I am lost in my head above the Earth; I don't know which way is down. I am a flattened human oscillator, the spins inside a spinning machine, a miserable pudknocker. I have lost Burbank again. I have lost the Earth, lost my way. Maybe hell is up and not down. I think I am losing my mind . . .

My sunglasses fall, lenses bouncing all over the cockpit.

"What the hell was that? Over," says Hollywood, but I can't answer him. I'm gone now, in a crushing vortex of nausea. More than being shot down, what I don't want to do is throw up—especially with that camera on me. I close my eyes, and all I can see is the footage: a kaleidoscope of vomit, cursing and broken aviator glasses. And meanwhile, Burbank is killing me. He's shot me twice already, and now he's back on my tail. For a brief, spinning moment, I consider giving up drinking. But heroes don't give up.

I open my eyes and bear down. I climb and climb, diving straight up, away from the Earth. I invert, twist and drop, and now I'm behind him—that skinny little freelancer—and I'm shooting him up.

"About time," says Hollywood. "Over."

"Fuck you," I mutter, but muttering is just speaking when you're in a cockpit. "Over," I say. Then I lock on Burbank one more time.

Next hit wins. But I really don't care about winning—which is rare for me. I just want to be back on the ground. I want to kneel. I want to repent. And in the time it takes me to think this, Burbank hits me again.

"Toxic down," says Hollywood, over the open channel, and he sounds downright cheerful. I fly through the air, straight as an arrow, then dipping each wing, trying to level my guts and my brain. Las Vegas shimmers on the horizon. In the sky, I see vultures and burning cacti before Hollywood regains control.

GOING DOWN DRINKING

In picturing ancient Rome, you might come up with something that looks a lot like Vegas: a crass, orgiastic place full of scantily clad women, corpulent douchebags and giant cups of booze. To quote the Roman historian Columella: "We spend our nights in licentiousness and drunkenness, our days in gaming or sleeping, and account ourselves blessed by fortune that we behold neither the rising of the sun nor its setting."

But before Rome became the original Sin City, it was, for two hundred years, an officious dry state—more a vast, ruthless desert than a debauched oasis—where devotees of Dionysus were treated with suspicion, then sanctioned, then hunted down and slaughtered by the thousands. This was, in many ways, civilization's first attempt at widespread Prohibition, and eventually it conjured up the same forces that it does to this day: corruption, bingeing, madness and a whole new form of hangover.

Rome's shift from dry to wet to drowning in the stuff was due to its early successes. As the empire grew, so did its armies—spreading out farther and farther. And despite an early sense of austerity, those Romans knew the importance of booze when it came to battle. More victories meant more war, which meant more wine, until after a while—and no doubt with a bit of lobbying from the winemakers—it started flowing over into the cups of the upper class. And then, like a teenager finally getting hold of a case of beer, the insatiable empire soon did away with all those pesky Platonic ideals of *symposium*—moderation, balance and rationality—and the age of binge drinking began.

Pliny the Elder described the winemakers of Pompeii cooking themselves in the public baths while glugging their own booze: "Still naked and gasping they seize hold of a huge jar . . . and as if to demonstrate their strength pour down the entire contents . . . vomit it up again immediately, and then drink another jar. This they repeat two or three times over, as if they were born to waste wine and as if wine could be disposed of only through the agency of the human body."

In terms of the physical hangover, there are a couple of ways to look at such behavior. There are those who believe that sweating it out in whirlpools and saunas will help rid your body of toxins, and also that nothing clears your system like reverse peristalsis. But then again, overheating your body will add to dehydration, and drinking more after you've vomited is a great way to get full-on alcohol poisoning.

But whether such methods of drinking increased or decreased the physical aspects of the hangover, they surely played havoc on the metaphysical ones. As Pliny wrote of the morning after: "Next day the breath reeks of the wine-cask, and everything is forgotten—the memory is dead. This is what they call 'snatching life as it comes!' when, whereas other men daily lose their yesterdays, these people lose tomorrow also."

Weirdly, vomiting was all the rage in ancient Rome—even among teetotalers. The first Caesar, Augustus, induced vomiting to avoid drunkenness if the occasion called for more than a pint of wine. By the reign of the third Caesar, however, the throne had become a drunken perch of madness.

As he rose to contender for the rule of Rome, Mark Antony became pure, boozed-up, burning mojo—bedding Cleopatra by night, throwing up on his own sandals in the morning, shaking it off, then leading his armies into war. In his final battle, he too was dressed as Dionysus.

The sadistic, drunken Caligula made people watch as he had sex with their loved ones, then declared his horse his main consul. The next Caesar's issues were so prolific that psychiatrists to this day are still trying to figure him out. Wrote Dr. Frances R. Frankenburg in a 2006 case study: "I, Claudius, am paranoid, hypomanic, habitually drunk, and have severe abdominal pains. My family is dysfunctional, and my wife is trying to kill me. What's my problem?"

His problem, as the doctor sees it, is one of mental and physical illness caused, or at least exacerbated, by both binge drinking and lead poisoning. She recommends lithium, psychotherapy and "education about healthy dieting and counseling against high-risk behaviors associated with alcoholism."

But really, since the Roman ruling class tended to favor fancy lead goblets forever filled by servants and slaves, Claudius's issues were also those of the whole empire. And by the time Nero, in the guise of both priestess and bride, married himself to one of his knights, the air itself in Rome must have felt like a combustible mix of madness, blasphemy and alcohol.

When it finally went up in flames, Nero embarked on one of the most famous binges in history, drunkenly fiddling as everything around him burned—until the sun rose on the smoldering ruins of the morning after. But of course, that tends to be the way: a helluva lot of booze to build an empire—then just a bit more to bring it down.

WHAT HAPPENS ABOVE VEGAS
(WHEN YOU JUMP OFF A VERY TALL BUILDING)

I am standing on the highest balcony in the United States, looking down on the lights of Las Vegas at night, about to jump, the wind blowing so hard, it's all I can hear. If I'm thinking anything, I'm thinking about fear.

This dive off the Stratosphere Tower is called "a controlled free fall," an oxymoron of very specific proportions, worked out to the last scariest inch. You launch out from the platform with a harness and cables attached to you. Then you fall—and keep on falling, until, at some point, you start to slow down. Theoretically, it's possible to land on your feet. It is the highest of these jumps anywhere in the world.

On the way up here, there were five of us in the elevator, including the current *Playboy* Playmate of the Year. Her PR team had decided it would be good publicity for her to jump off a building with some reporters. If you think elevators are generally awkward, imagine that one, which just kept rising to a place none of us wanted to go.

But while my colleagues were trying to make breezy conversation, I was still queasy from this morning's dogfight, and my head was far too full—of drinking, driving and flying; of writing, reading and interviewing; of my girlfriend, who wants to get married; my dad, who's scared of heights; my baby boy, who thinks he can fly—but first of Levi Presley.

In 2002, at the age of sixteen, Levi Walton Presley scaled two fences on the 109th floor of the Stratosphere and leapt to his death. It was 6:01:43 p.m. when he jumped, and 6:01:52 when he landed. I've read far too many accounts like this—mostly because I spent six years working on a novel about a guy who ghostwrites suicide letters. But it wasn't until the elevator ride up here that I suddenly put it together: that I was going to be jumping from the same spot that Levi did. It made me think of my own lesser losses—of never quite experiencing the life you're living; of doing things you dream about, like driving race cars and flying fighter planes, but all while feeling too fucked up to really, fully *be* there.

Around floor 60 I tried to gauge my presence, my *there*ness, but couldn't quite do it. Maybe it was all that g-force, both yesterday and today; the 100 percent oxygen; the pills and chews and other things; being unable to reach my boy on the phone; the guns and dryness; or—who the hell knows—maybe even the booze.

And then something happened. Right around floor 85, the desperate banter stalled. The small talk ran out. There was a moment of silence. Then the heroes started to mutter: "I'll go first," said Iowa Crayon at about the same moment as the California Kid. "What ever happened to 'ladies first'?" said the Playmate.

Nobody spoke again. The elevator stopped. The doors opened. And the Earth was a world below us. We watched two strangers—newlyweds—jump, one after the other. The man looked back, staring right through us. And here's what we could all divine: it didn't matter who went first; somehow or other, we'd override our instincts and make the leap. But to be the *last*—to watch the others jump, the fear building, then be up here alone—that would be a final, unnecessary terror, and the only way I'd be sure to feel something.

"I'll go last," I said.

So now I am here, alone with the fear. And it's hitting me like these rare desert winds. There's a guy with a harness, a headset and a long hooked pole. He's trying to get a hold of that wire whipping about in the dark air. And until he does, I'm not attached to a goddamn thing. The others are all down there on the ground. Elated, traumatized, injured, giggling, dead—I have no idea. The giant casinos, more than a hundred storeys below, are the size of Monopoly hotels. I can block one out with my thumb.

The wind is hitting gale force now, and the guy is bellowing into his headset—the one thing you don't want to hear before you leap off the highest tower west of the Mississippi: "Last jump! No more after this! This guy will be our last!"

Fighting against the wind makes one moment seem like a thousand, and now I'm thinking again: of my boy and Levi Presley, of the King of Rock 'n' Roll and Kingsley Amis, of marriage and death, of never being drunk and never being sober, of being the last jumper and *being the last jumper*, of being committed and *being committed*. I am 855 feet above the ground, and now I'm just trying to breathe.

Finally, the guy with the hook gets a hold of the cable. He's off-balance as he turns toward me, and I toward him, not holding onto anything. He gets me fastened and steps back, and now

I'm shaking in the wind on the lip of the world—every fiber of my being trying to stop me from jumping. But that, of course, is just instinct—a silly fear of heights and/or death. This acutely engineered fall is probably safer than driving to the grocery store. I flex my calves, curl my toes, taste my fear—and some leftover bile—and look out at Las Vegas. My mind has slowed to three distinct thoughts:

1. Bless your poor heart, Levi.
2. How about this one, Kingsley?
3. Life is a helluva thing to try.

And then I jump.

IF YOU REALLY push off with your legs, stretching out into the wind, you rise for an instant before you start to fall. And in that moment, everything happens. Your brain, in terror, reaches out for something, somersaults through nothing, everything, then back into your body, which is freaking out and trying to recalibrate . . .

And now, you are just there. More than you have ever been. Falling and flying at the same time, into the lights of Vegas.

And this is what you see: your life flashing before you; the *Playboy* model on a poster ten stories high on the side of a building, like a giant, ascending ghost; *Hangover 3* on a billboard behind her; and the world hurtling upward. You are screaming and laughing and going to be fine. There is no way you'll stop in time, but your body and brain are past all that; you are one with the brilliant descent.

Maybe this is where Kingsley was going with that whole mine shaft/open cockpit thing—a theory that (as with hiccups) the right kind of shock will blast a hangover right out of you. And perhaps there's something to be said for it. A massive adrenaline spike can outweigh most physical realities of the human body, especially when combined with a fight-or-flight scenario. But to not just crash once the adrenaline has run its course no doubt requires something else—something shocking enough, both physically and metaphysically, to somehow reboot your system.

I hit feetfirst, bouncing off the ground, screaming and laughing and howling.

And just like that, I've been rebooted. I can tell, because now all I want is dinner and drinks.

This brings up a vital point: How do you know when the hangover is done? It's tempting to put the answer in esoteric terms, like the way we talk about being in love: you know when you know; you know when you no longer ask. But I've come to believe that it's simpler than that, and more akin to heartbreak: your hangover is done, truly over, when you're ready to start on another one.

By those terms, jumping off the Stratosphere has cleared all vestiges of my lingering hangover. And now, as we ascend once more to a rooftop Vegas restaurant, I feel high and hungry and invincible. I order half the menu, then open up the wine list. It shimmers in the light as I look for a place to land.

THE MALEFICENT SEVEN

Since early antiquity, and for unclear reasons, numerous orators have devised lists of the kind of drunks that people can be. The Elizabethan actor Thomas Nash, in his *Pierce Penniless's Supplication to the Devil*, riffed on ancient correlations between man and beast to come up with a list of eight main types: *ape drunke*, who always has a pretty good time; *lion drunke*, who is fairly aggressive; *swine drunke*, a sloppy mess; *sheepe drunke*, a big know-it-all; *mawdlen drunke*—not an animal, it seems, just a particularly maudlin person; *Martin drunke*, some guy named Martin who drinks for so long he becomes sober again; *goat drunke*, a total letch; and *fox drunke*, whom you'd be a fool to trust.

Based on two decades of personal experience now morphed into research, I'm pretty sure there are just as many ways to be hungover. It is one of the many things that make writing about hangovers nearly as slippery as trying to treat them. So, with that in mind, and a mind somewhat addled by research, I have attempted to corral the hangover into meaningful categories. I've named them the Maleficent Seven—even though three are actually quite positive.

Most real-life hangovers—like a queasy bull rider scrambling out of his chute, or a grasping writer drunk on metaphors—will likely straddle a couple of these categories at once, then tumble over others before crashing into sawdust. But there it is: the imbalanced nature of corrective chemistry, purposeful lists and every kind of rodeo. And so, without any more stumbling around, here they are: the Maleficent Seven, along with some possible treatments.

1. THE CREEPER

Even if you've never had a drink in your life, you should recognize this one from the first chapter. The Creeper watches you from the corner of the room as you wake up feeling strangely okay. It follows you through your morning, just waiting for the perfect moment (riding backward in a limo, for instance), and then it strikes, jumping onto your back, teeth through your kidneys, its long tail wrapped around and snaking down your throat. You gag and shake, trying to wrestle free, but that just tightens its grip. Really, you've got to give in. Go limp, clear your thoughts and your bowels, drink some coconut water and crawl back into bed. Once you're asleep again, it should creep off somewhere else.

2. THE BLISTERING BARNACLES

In dubious honor of the heavy-drinking Captain Haddock (compatriot of Tintin, that oddly teetotaling mid-century journalist), the Blistering Barnacles is an ongoing edginess—there when you wake up, then stuck to you all day. Even if you've slept, you're not at all rested. You're worn out and crabby, with a festering feeling. Though brave and big-hearted, Haddock could be a bit ornery, especially in the mornings. "Billions of blue blistering barnacles!" he'd bellow upon stubbing his toe or taking an accidental sip of water, spitting it out to wash down the taste with rum. Try paring back the Blistering Barnacles with a plate of pickled herring, and then the start of a new adventure.

3. THE TROOPER

So named for both the drinker and the malady, this is what you get when you rise to the challenge of a challenging hangover. It could be as simple, yet onerous as pushing through a day at work, the effort itself creating an ever-increasing string of new symptoms. Or maybe, like John McClane in *Die Hard: With a Vengeance*, your drinking binge has been interrupted by a mad bomber with a German accent, things keep blowing up, and you can't even find an Aspirin. Or if you're David "Boomer" Wells, you pitch the fifteenth perfect game in Major League Baseball history. The Trooper requires a sense of mission, along with a feeling of the impossible. It may also require amphetamines.

4. THE BURROWER

The Burrower sucks. And also it burrows. While the Blistering Barnacles stay close to the surface, jangling at your nerves, the Burrower digs right in. And the longer it's there, the deeper it goes—into your head, heart, stomach and soul. A chemical, spiritual parasite, it sucks at your very being, hollowing you out piece by piece. And you don't want to give this one too many hairs of the dog; the Burrower thrives on next-day booze, kicking back and drinking it in, gaining strength as you grow weak. In fact, once it's there, there's not a lot you can do—except maybe jump off the Stratosphere. The Burrower likes heights about as much as it likes adrenaline. Shock the host, and you might just shake out the bug.

5. THE JOHNNY FEVER FEVER

This impressive, endless form of hangover is named for the most consistently strung-out character in modern history. Over the course of *WKRP in Cincinnati*'s eighty-eight episodes, disc jockey Dr. Johnny Fever was known for two things: never playing the hits and always being hungover. The Johnny Fever Fever is, by its very nature, only attainable by a truly good-hearted drunk (not to be confused with a self-pitying alcoholic) and is as much a badge of honor as an eternal curse. To have it requires fulfilling a certain destiny while never becoming fully sober. Keepers of the Fever include Winston Churchill, Charles Bukowski and Keith Richards. The Johnny Fever Fever requires the hair of the dog—not to cure it, but to keep it going.

6. THE SHINING

Upon waking, you feel slightly aglow—a low-grade dulling of your senses that leaves you paradoxically open to the whims of the world. An unlikely gift to artists, philosophers and inventors, the Shining is a sort of booze-soaked muse that is hard to conjure and then, when it appears, is often ignored. The trick is to drink enough, but not too much, and then have nothing to do in the morning. No one knows quite why it works, but if it does, you'll become a sort of pickled divining rod for all these little bits of inspiration floating around in the hazy light of day. Those who learn to recognize and even replicate the sensation will often get drunk at night, with passionate precision, in hopes of artistic revelation when they wake. This is why some of the heaviest-drinking scribes, from Hemingway to Hitchens, Dorothy Parker to my dad, are most counterintuitively productive first thing in the morning. I can't think of a reason to cure this type of hangover.

7. THE COMPLETE DEVASTATION

This is the big one—the only known example of self-inflicted, semi-premeditated self-torture that exists in nature. The Complete Devastation is when you feel like you're dying, even though you're not. It's when you wish for death, even if you want to live. It's not the Devastation (versions of which you may have experienced, but which don't require their own category—each being, essentially, "a really bad hangover"). The Complete Devastation combines the worst of physical and metaphysical hangovers with ultimate fallibility, intrinsic collapse and pure cosmic chaos. The Devastation involves urine and vomit. The Complete Devastation trades in blood, feces and spinning souls, and often results in the following: hospital, jail, the army, organized crime, religious cults, interventions involving in-laws or sometimes even rehab. The Complete Devastation can change the direction of your life. And also not. If you've lost track of your Complete Devastations, I wish you the best—and maybe a bout of Johnny Fever Fever.

WHAT HAPPENS ABOVE VEGAS (WHEN THE DOCTOR ARRIVES)

I am crashed on the gargantuan purple couch in my upgraded suite overlooking the most nauseating roller coaster in the world, waiting on the hangover doctor. I have seen his photos, read what he's written, interviewed his colleagues at Hangover

Heaven and even talked to him on the phone, but I'm still not sure what to expect.

"Should we have to lose an entire day of our vacation because the bartender over-served us the night before?" asks Dr. Burke on his website. "I say NO. With my treatment protocol, I can take you from a semi-conscious, porcelain-hugging, hit-by-a-truck hangover to feeling like you're ready to take on the world in less than 45 minutes. I think this is a major development in medicine and solves a significant problem for people that like to party and have a good time . . . especially here in Las Vegas."

The accompanying photos look like a TV doctor's publicity shots—or, more specifically, the headshots of a champion-surfer-turned-actor auditioning for the part of Dr. Burke on a soap opera. His long blond hair, much like his prose, might be purposefully ironic or stunningly earnest; it's very hard to tell. When asked, the people who work for him describe him as "a genius," "some kind of genius" and "sort of like a genius." It would not surprise me to discover he belongs to the Church of Scientology.

As Dr. Burke's own press puts it: "He is the first physician in the United States to formally dedicate his career to the study of veisalgia, the medical term for the common hangover . . . The field of medicine has done a generally poor job of addressing veisalgia and it is time to end this scourge."

Veisalgia is a somewhat recent doctorish word, from the Norwegian *kveis*, meaning "uneasiness after debauchery." But what strikes me deeply, lying on this purple couch with my eyes screwed shut, is the word *scourge*. It is practically onomatopoeic. Forget hangover, we should call it the scourge! *How are you feeling this morning? Ohhh, a bit scourgey . . .*

Then comes a knock on the door.

DR. BURKE, it turns out, is all I could have hoped for, and more. His spotless scrubs are the color of my couch, his golden locks a He-Man halo. He has a bachelor's degree in Classical Studies. He is earnest, yet easygoing, neither ironic nor jokey, and only the slightest bit admonishing.

"Your intake last time was substantial," he says, setting up the IV stand, his voice like Virginia ice melting in a tumbler, his handsome face so distractingly symmetrical. "Had you waited the requisite amount of time to be hungover, rather than intoxicated, it still probably would have taken two bags, rather than one—which is all we had time for, given your schedule."

"Sorry," I say. And truly I am. Sorry for my unshaven cheeks, my bloodshot eyes, my mismatched socks, my writer's physique, my poor earning potential, my lack of discipline, my bad habits and, especially, the asymmetry of my stupid, sorry face.

"How are you feeling now?" says Dr. Burke.

"Hungover." The word comes out like a white flag unfurling. Perhaps this is how his magic works: by his very presence—his very well-groomed, very fit, very healthful, very, very presence— he makes you feel, in contrast, the depth of your debauchery. And so there is nowhere to go but up.

"How much did you drink last night?"

"About half as much as last time," I say, though really I can't be sure.

He pulls two bags from his suitcase on wheels and starts hooking up the first. It's the same as I had before: Myers' cocktail. "We'll try to go through both of them," he says, puncturing my arm. He gets the bag going, then starts to ready the oxygen.

There's nothing judgmental in his aspect, but I feel like I should stick up for myself. "Surely I'm not the first," I say, "to show up more drunk than hungover?"

"I suppose not," says Dr. Burke, placing the mask over my face. And then he starts a story, in that deliberate Duke drawl of his. A guy calls up, and he's frantic and slurring. His buddies are in the slammer, they got too drunk, and he's too sick to go get them.

"And just like that, it occurs to me," says Dr. Burke. "The way things sometimes do. And I say to him, 'Now, sir, are you driving?'"

And yes, he is. So the doctor directs him right into a parking lot and whispers him down to sleep. "Call when you wake up," he says. So that's what the guy does. They bring him in, sort him out.

"Made him right as rain," says Dr. Burke, "and nobody got hurt."

Despite the relevant bits about drunk driving, the story might not be directed at me at all. The mask is snug, the oxygen flowing, and I'm just trying to do my job—though it'd be fair to ask, right about now, what that even is. The business of being a freelance drinker/hangover author is, at the very least, a dubious one. And the obstacles are exacting. For example:

THE HANGOVER ITSELF. No matter the purpose, or what is going on, it is hard to generate the sort of gumption and wherewithal to press, dig or challenge points to a possibly meaningful level when you are this hungover.

THE OXYGEN MASK. Interviewing people when you're hungover is always a drag. Caring is a trial, and trying to focus is like swallowing bugs; a plastic cone over your face doesn't make the process any easier.

THE ONE-AND-ONLY HANGOVER DOCTOR. He is a wonder: confident and comfortable in his own symmetric skin, so shiny clean that it's fair to wonder: Has *he* ever been hungover? I pull back my mask and ask him.

"I loved red wine and I hated red wine," says Dr. Burke—a haiku in the making. "I was the president of the Las Vegas Bordeaux Society, and kept on waking with a headache in the morning. . . . That's when I knew: something had to be done."

Since the Bordeaux-inspired inception of Dr. Burke's business, the hangover industry has expanded like blood vessels in a boozy eyeball. The doctor attributes this to the trifecta of American influence: economic fear, health obsession and Hollywood. In this instance, the perfect storm came in the swirling shape of two studies and a movie. One study said that hangovers cost the US economy $150 billion a year. Another study suggested that herbs and extracts, particularly prickly pear, might alleviate some hangover symptoms. And then *The Hangover* became the top-grossing R-rated comedy of all time. Within a year, every convenience store in North America was overflowing with little-bottle cures.

"The Hangover movies have been good for Las Vegas, and great for this industry," says Dr. Burke, as he hooks up the second bag. "They say this third one is the last, which is too bad. But I'm sure people will still get hangovers."

I nod in agreement. Dr. Burke gets the drip going, then starts on another story. They treated this guy's hangover, but he kept complaining of a gagging feeling. He had a history of severe acid reflux, so Burke, to be safe, sent him to the ER. There, they did an esophagogastroduodenoscopy (as they're wont to do), and found a piece of steak lodged in his gullet. But that's not all. Apparently, while living in some far-off land, the guy had gone to another doctor a year earlier with the same complaint. And this other doctor, misdiagnosing a throat full of rib eye as terminal cancer, had given him three months to live. He'd been drunk ever since.

"It just goes to show," says Dr. Burke, shaking his perfect head.

"It sure does," I say, feeling altogether agreeable. There are probably more questions I should ask, but I'd rather just drift for a while, amid all this oxygen and vitamins and electrolytes and soft, plush light. I know I've got to go climb a mountain in the Nevada heat soon, and then zip down it—and *then* fly a helicopter, and later get drunk again. After three days in Sin City, I am worn out and sleep-deprived. But all in all, at this very moment, I'm feeling almost alright. I'll give the doctor that.

I lie back on the purple couch, close my eyes and breathe.

PLENTY OF AVERSION: A VERSION OF PLINY

My father, a man of extremes, tends to dash back and forth between rampant hedonism and Spartan self-punishment with barely a stop in the middle. He is a prolific imbiber, and he used to be a two-pack-a-day smoker. For years, the longest he went without a cigarette was the four and a half hours it took to run his first marathon. But to his credit, he spent most of my childhood trying to quit.

He tried patches and gums, gurus and swamis, and eventually someone convinced him to pick up The Jar. Half full of brown, sludgy water and cigarette butts, he carried it wherever he went. I even remember a cord that held it like a giant sloshing amulet around his neck. Every time he felt like smoking, he'd open The Jar, hold it up to his face and inhale the batch of cigarette soup. Only then was he permitted to light up, smoke, then drop the new butt in.

"Aversion therapy!" he'd say brightly, swirling The Jar to help it thicken as my sisters and I watched, gagging and horrified. Although he swears he's never had a hangover, the only days The Jar really seemed to bother him were when he'd drunk more than usual the night before. That was a level of therapy that even my dad's love for cigarettes could just barely handle.

He finally did quit smoking—though I don't ever remember him trying to quit drinking. In my father's day, the hangover—whether or not he actually experienced one—was recognized, for the most part, as a necessary obstacle. It could also be seen as nature's great aversion therapy.

In the *Journal of the American Medical Association*, Dr. Michael M. Miller describes hypnosis as a treatment for alcoholics: "In most instances I simply intensify the disgust, aversion, and displeasure of the hangover state by having the patient actually relive one of his worst hangovers."

But of course, there is a world of difference between hangover remedy and aversion therapy. One tries to help the patient out of the state of hangover, while the other uses the hangover against the sufferer, amplifying symptoms in an attempt to solve a presumed underlying pathology. And yet it is a difference that seems to have been either accidentally or pathologically confused by the very thin ranks of modern-day hangover histories. I am referring here to Clement Freud's *Hangovers* (1980), Keith Floyd's *Floyd on Hangovers* (1990) and Andy Toper's *The Wrath of Grapes—or the Hangover Companion* (1996).

Each of these slender, somewhat informative and entertaining volumes refers back to Pliny the Elder as the first, most complete compiler of hangover cures. But then, Pliny was really the first, most complete compiler of *everything*. His *Naturalis Historia* is generally viewed as the original encyclopedia of all human knowledge, addressing everything from the movements of the planets to insect mating habits to every kind of hooch in existence.

And apparently that's what it took—an obsessive drive to find and record every facet of the known universe—for the hangover to briefly come into view. Though not a drinker himself, Pliny cataloged the aftereffects of alcohol with the same ruthlessness he brought to everything else: "Tippling brings a pale face and hanging cheeks, sore eyes, shaky hands that spill the contents of vessels when they are full, and the condign punishment of haunted sleep and restless nights."

To this day, it is hard to find any historical study of the subject that doesn't make reference to Pliny, especially when it comes to possible cures. In his chapter on ancient remedies, Clement Freud offers this list:

- eggs of a night-owl in wine (Pliny);
- a mullet killed in red wine (Pliny);
- two eels, suffocated in wine (Pliny).

A decade later, Floyd digs in a little deeper: "[Pliny] believed that prevention was better than cure and that the wearing of purple robes and the use of amethyst-studded drinking goblets would counter the fumes caused by the wine. . . . Pliny recommended, if on waking one is suffering, some lightly boiled owl's eggs for breakfast. However, if the hangover persists, he suggested one should partake of a dish of stewed eels."

Then, with a vague kind of precision, Toper sums it all up: "In one of his books [Pliny] states that hangovers can be avoided by wearing a necklace of parsley when retiring to bed after a heavy drinking session; or cured the following morning by swallowing two raw owl's eggs in wine."

But to find these things in "one of his books" is no easy feat, considering estimates put Pliny's output around 160 complete volumes of study. And as it turns out, none of those curative suggestions attributed to him appear to exist; in fact, just the opposite.

In regards to purple robes and amethyst, this is what Pliny actually wrote: "The Magi falsely claim that the amethyst prevents drunkenness." And rather than being hangover remedies, all those dead sea creatures in wine were in fact prescribed as aversion treatments—a sort of ancient rendition of my father's murky jar, but for drinking rather than smoking: "Red mullet killed in wine, or the fish rubellio, or two eels, also a sea grape rotted in wine, brings distaste for wine to those who have drunk of the liquor." And the same with owl's eggs: "The eggs of an owl for three days in wine given to drunkards, produce distaste for it."

But in a sort of journalistic broken telephone, every one of our few hangover historians seems to have skipped right over this part of the equation, so that Pliny's "cures" have become a popular truism. Even the brilliant Barbara Holland states, "The Roman sage Pliny the Elder recommended two raw owl's eggs, taken neat the next morning." And yet there is no evidence of this, whatsoever.

Only Clement Freud seems to have got it partly right, explaining Pliny's "cures" in an inverse wino sort of way: "They were meant to be taken in wine. It would be more accurate to describe them as distinctly flavoured hairs of the dog."

PART THREE

THE HAIR THAT WAGS THE DOG

IN WHICH OUR MAN AT HOME FALLS ILL, BECOMES A MUTANT
AND STARTS THE NEW YEAR BY PLUNGING INTO FREEZING
WATERS WITH ONE HUNDRED HUNGOVER POLAR BEARS.
APPEARANCES BY HIPPOCRATES, A COUPLE OF FAMILY
DOCTORS AND THREE WISE MEN.

*The best thing I have done is drink . . . I have written
much less than most who write, but I have drunk
much more than most who drink.*
—GUY DEBORD

M Y FAMILY DOCTOR IS LAUGHING AT ME. I THINK it's the first time I've ever heard his laugh: a sort of dry Canadian chuckle, the sound of genuine amusement topped with a high note of irony. We are discussing the possibility that I have somehow acquired an allergy to alcohol.

"Take me through it again," he says, leaning forward, with a sense more of humor than concern. I seem to be the only one worried here.

"It happened first with the mimosas," I say. "At breakfast . . ."

We were on vacation, my girlfriend, Laura, and I. Nothing extravagant—just a drive to a small town for a few days. We golfed and ate and drank, and I ate and drank a lot. We were staying at an inn, and the day we were to leave, they had a big buffet brunch, mimosas included. I don't usually drink much orange juice, but mixed with champagne and a hangover, it's kind of hard to resist. Then, halfway through brunch, something started to happen.

"Are you alright?" asked Laura.

I tried to answer, but my mouth felt puffy, and so did my head. And my head also felt hot, and now I was sweating. I took another sip of mimosa and the room began to spin. Then I switched to water. That's when we knew something was wrong.

About a quart of water and an hour later, the symptoms had dissipated enough that I tried driving back to the city. But partway down the lakeshore road, Laura asked me to stop in a town so she could buy Tylenol, a thermometer and another quart of water. I felt like I was burning up, but I had no fever. My face in the rearview mirror was bright red. And by the time we got back to the city, there were welts on my arms.

I didn't drink the rest of the day. I cooled down and felt a bit better, but the next evening we went to a dinner party, and after just one beer it happened again. I soldiered on with a couple glasses of wine and a hot, red face. Then it happened once more the day after that. That's when I stopped drinking altogether and phoned the doctor.

"SO, HOW LONG ago was that?" he asks, studying the marks on my arm.

"Three days," I say. "It's my *job* to drink. My book depends on it." But I've told him this before. Not just about this one, but other books and other jobs—and without these symptoms and blotchy arms. "I *can't* be allergic to alcohol!"

The doc leans back. "I don't think it'd be that, exactly. People don't really get *allergic* to alcohol. Certain drinks, maybe, because of other things in them . . . but it sounds like you're talking about all different sorts of drinks."

"I am! That's exactly what I'm talking about."

"With the same negative effects . . ."

"That's precisely right! So, what should I do?"

"What do *you* think you should do?"

I feel like I've tripped into a trap.

"Umm . . . can I maybe see an allergenist?"

"*Allergist.*" (With every new word, I am failing—as both writer *and* drinker. *What will be left of me when I leave this place?*) "I thought you didn't want it to be an allergy."

"But to something *else*, maybe?"

"And until then?" he says, reaching for a pad of paper.

"What do you mean?" I know exactly what he means. It'll be at least a month before I see an allergy guy. It's always at least a month.

"You should stay off the booze," he says. "At least until we figure it out. Drink some water instead, okay?"

"Sure," I say, looking down at my feet. Walking out of his office, they're all I can see of myself.

MAN DRINKS DOG

Throughout history, people have done all sorts of weird things involving animals in the hopes of curing a hangover. It is said that the epic drinkers of Outer Mongolia pickled the eyeballs of sheep, horse wranglers in the Wild West made tea out of rabbit shit, and my Welsh ancestors roasted the lungs of a pig—all very literally. But the most common remedy has always been figurative: to pluck a hair of the dog that bit you.

That catchy metaphor goes back to at least 400 BCE, when Antiphanes wrote these words (or at least their equivalent in ancient Greek). And they appear to be riffing on something even earlier:

Take the hair, it is well written,
Of the dog by which you're bitten,
Work off one wine by his brother,
One labor with another.

Thanks to Antiphanes's contemporary Hippocrates, the forefather of both the Hippocratic oath and homeopathy, it was a concept popular to all forms of medical treatment: fight fire with fire, yet do no harm. A tricky combination, but one that people have been attempting forever. And of course, it wasn't just hangovers that were treated with alcohol. Hippocrates devised an elaborate system of wine therapy, prescribing different types for different ailments and incorporating it into the regimen for almost all chronic and acute illness.

Galen the Greek, who served in the court of Marcus Aurelius, brought Hippocrates's methods to the Roman Empire. Prolific as Pliny, he published two and a half million words in his life-time, much of it about wine therapy. Amidst a hundred other wine-soaked remedies, he used it to treat the wounds of gladi-ators, and apparently not one of them died from infection—from decapitation, disembowelment and hungry lions, sure, but not from infection.

By the eleventh century, medieval healers were still looking to Hippocrates, Galen and other ancient doctors whose teachings had been translated by monks into Latin and compiled into the great medical tome *Regimen Sanitatis Salernitanum*. This com-pendium of all medical knowledge prescribed wine and other alcoholic beverages for everything from indigestion to insanity. And it tells you exactly what to do should you exceed the recom-mended dose. "If you develop a hangover from drinking at night, drink again in the morning. It will be your best medicine."

And of course, booze was the least questionable ingredient in many curative elixirs. The *Dispensatorium Pharmacorum,* a medical dictionary from the mid-sixteenth century, contains recipes that combine wine with ingredients such as the ashes of scorpions, dog excrement and wolf's liver. And an article from *The London Distiller* in 1667 explains how to make an appar-ently well-known tonic from a crushed human skull: "Take the Cranium-Humanum as you please, break it into small pieces . . . then put string fire to it by degrees, continuing until you see no more fumes come forth; and you shall have a yellowish spirit, a red Oyl, and a volatile salt." The resulting liquor is supposed to help with "the falling-sickness, Gout, Dropsie, infirm Stomach; and indeed strengthens all weak parts, and openeth all obstructions, and is a kind of Panacea."

If, however, your hangovers tend to be accompanied by tinges of remorse, you might want to avoid getting drunk on crushed human skull. But then again, there's always this for the morning after, from Andy Toper's list of ancient hangover cures: "In Old Europe it was widespread to cultivate moss inside a skull, dry it, powder it, then snort it"—a bit of the skull that bit you.

It is also possible that, at certain times and places in history, whole towns, cities and civilizations never got a chance to be hungover—they just kept a slight buzz going all day, then into the next, apparently without a lick of guilt.

Today, it is this aspect—the degree to which your hangover makes you feel guilty—that may decide what kind of hair of the dog you choose: a warm, stale beer, a spicy Bloody Caesar or one of countless cocktails invented specifically for this purpose. Such concoctions tend to come in two overarching categories: sweet, soft soothers meant to ease you back to baseline with something milky, fruity, relaxing and restorative (these have names like Morning Glory, Milk of Human Kindness and Mother's Little Helper), or short, sharp shocks intended to twist your system into sobriety with bitterness, blinding heat and/or gag reflex (Khan's Curse, Suffering Bastard, Guy Fawkes's Explosion . . .).

Historically, this second variety could contain anything from anchovies to ammonia, garlic to gunpowder. As Clement Freud (sounding more than a little like his overly analytical uncle) puts it, such tonics also "sublimate any guilt feelings" by subjecting an already remorseful bastard to a more precise moment of suffering: "Liquid cures of this type," he opines, "owe much of their effectiveness to the popular belief that anything that tastes really disgusting must actually do you good."

But guilt feelings aside, does a hair of the dog make practical sense? Sure it does, and always has. Even the National Institutes of Health in the US admit that "the observation that alcohol readministration alleviates the unpleasantness of both AW (alcohol withdrawal) and hangovers suggest that the two experiences share a common process."

But in regards to "readministration," those in such institutes also warn you should never, ever do it, lest you turn into a certified drunk. In 2009, Dutch researcher Dr. Joris Verster published a paper addressing precisely this. "The 'Hair of the Dog': A Useful Hangover Remedy or a Predictor of Future Problem Drinking?" was based on a survey of Dutch undergrads. It revealed, among other things, that those who used alcohol as a morning-after treatment consumed approximately three times as much alcohol, and that those who did so more often had "a significantly higher lifetime alcohol dependence diagnosis." But of course, there's an aspect of chicken-and-egginess to all of this. And also, Verster's paper never did tackle the first part of the question: "A Useful Hangover Remedy?"

The hair of the dog may, in fact, work in ways we never imagined. In *Proof*, Adam Rogers's excellent recent book about the science of alcohol, the author suggests that "ethanol might help with a hangover because it stops the body from breaking down methanol." As Rogers explains it, ethanol is the magical essence of alcohol, while methanol is a nasty molecule that, at low levels, sneaks into most alcoholic drinks and, at high levels, might just kill you. Broken down, it becomes the poison known as formaldehyde. While recognizing that some studies dismiss its effects, Rogers concludes that "one piece of evidence is suggestive: the relative efficacy of the 'hair of the dog'—drinking more booze."

THE HAIR THAT WAGS THE DOG
(THE DOG WHO WRITES THE BOOK)

It is more than a month later—a very long, very dry month—
and I am sitting in another office, showing another doctor the
marks on my arms. I have barely drunk for weeks, but still there
are blotches on my left forearm, and I feel, in general, like I've
been snorting moss from a dried-out human skull.

On my right arm are the marks this allergist put on me with
the pricks of a dozen laced needles: three rows of four, numbered
and underlined with a black felt pen. Some are little red dots,
while others are erupting volcanoes. These have been surrounded,
but not contained, by a blue circle of ink.

"Ragweed," says the doctor. I assume he says this a lot. "Pollen.
Dust mites. Dander." I am waiting for him to say "orange juice"
or at least some kind of citrus. Or ice cubes, little plastic swords,
roofies—*anything* that could explain what's been happening. But
now he's finished naming volcanoes, and . . . *nothing*.

"Did you check for oranges?"

He points to a small colorless dot.

"So, where does that leave us—assuming I haven't been drinking ragweed and smog?"

"Assuming that?" he says, face as straight as a dander-laden
needle. "Well, nowhere, really—at least regarding allergies. But
do you know what alcohol flush reaction is?"

"Sure," I say. "I think I do. But doesn't it only happen to
Asian people?"

"True, it *is* also known as Asian flush syndrome," says the
doctor—who himself appears to be of Asian descent. "It is a condition that affects about half of Asian people, whereby a specific
gene variant makes it much harder for the body to break down
something called acetaldehyde . . ."

Part of me actually knows he's talking about. After all, I've been thinking a lot about acetaldehyde lately—it being not only the chemical created when the body processes alcohol, but also commonly identified as the primary cause of hangovers. The condition he is describing is essentially one in which a single drink can cause immediate, severe symptoms.

"It can get rather nasty . . . not just blotches and rashes and such, but rapid heart rate, shortness of breath, sickness, migraines, mental confusion, blurred vision . . ." As he talks, it is like this doctor is pulling the air out of the room, into his own lungs, and I am gasping for one small breath of it. My head is pounding, my mouth is dry.

"Would you like some water?" he asks, but this seems like a fateful question.

"No!" I say. "I mean . . . yes. I mean . . . how can this be happening?"

The whole situation is starting to feel like some tone-deaf, obscure premise for an episode of *Punk'd* or *Candid Camera*: Asian doctor tells Caucasian hangover researcher that he has somehow acquired Asian flush syndrome, a genetic condition that causes instant hangover symptoms. The episode would be called "Turning Japanese." Or maybe "Drunkzilla."

He holds out a paper cone of water. I take a sip from this small inverted dunce cap and try to collect myself.

"I don't understand," I try again. "I've been drinking my whole life, so how could this be a genetic thing?"

The doctor shrugs. "Sometimes the body just makes sudden changes. They could be rooted in the brain, the nervous system. We don't always know how—but the body and mind have all sorts of ways to protect you."

"So, what do I do?"

"Well," says the doctor, "it happens when you consume alcohol . . ."

I am sick of doctors half-suggesting I know *exactly* what to do.

"*Other than* stop drinking?" I say.

"Well, niacin can exacerbate flushing," he says. "So I'd avoid that too."

I don't want to give him the satisfaction of admitting I don't know what niacin is. And anyway, I don't really care about what's exacerbating it. I care about what's *causing* it.

Other than booze, of course.

FOR MOST OF my drinking life, I've acted both proud of and sorry for my persistent resilience to alcohol. "The problem," I say, sloshing a glass around in my hand, "is that the problem is never big enough."

People get blackouts. They end up in hospital, or jail, miss deadlines, get fired, crash a car—or maybe become really sick, brutally hungover, lose days, just can't handle the drink anymore. But not me, *nosiree*. I can handle anything. So where's the reason to stop? Just keep on keeping on. That's the problem. And I figured that's how it would be until . . . what? Stomach ulcers? Gout? Sclerosis of the liver?

But *this*? Who's ever heard of this? It's enough to drive one to the murky depths of internet chat rooms—which is where I've discovered a medical forum dedicated to mutants like me. "Asian Flush but Caucasian?" is full of bewildered white drinkers who sometimes turn red. While these are not actual posts and people, these capture the spirit of various queries and complaints.

Whyismyfacesored reports:

It happens, but not EVERY time I drink. And it happens with all different kinds of drinks. My face gets red, with hot, itchy splotches. Very unattractive and embarrassing in public.

To which Roxbury2006 responds:

I thought I was the only one!!! I'm 30 years old and white. When I drink very heavily—or if I drink for 2 days in a row—that's when it usually happens. I can go to the pub on Friday, sink 10 pints of beer, a few Jager Bombs, a couple of whiskies and nothing. Then, the next day, I have a few pints, and suddenly it starts to happen. I've gone to doctors, had an endoscopy, blood tests, CT scans and nothing. A mystery. But I am determined to find the cause and will post on here once I do.

There has been nothing more from Roxbury2006, but I'm starting to notice a pattern—in that Asian Flush tends to happen after more than one day of drinking. LizJen99 writes:

This happens to me about a quarter of the time when I drink. And it doesn't matter what I'm drinking: cider, whisky, vodka, tequila, rum, wine, etc. (I'm not a beer drinker). . . . I've had it happen on the third consecutive day of drinking (but not the first two), and I've had it happen when I haven't had any alcohol for a few days and then have some. Like I said, only a quarter of the time.

BeatMan22, though he's had flushing symptoms since his first-ever drink, has become convinced that it's all in his mind:

This guy told me it was all mental, and then I realized that it was—because as soon as I told myself that, it didn't happen. But if I expected it to happen—then there it was! There are a lot of factors, for sure—including what you've eaten and where you are in your mind, mentally.

BootsAndKats is big on details. Once, when she had a banana daiquiri, with friends, in a hotel room, things went sideways:

I had so much chest pain I thought I was going to die. It lasted for a good 20 minutes and then finally went away. I found that drinking water slowly, and standing in 22F air temp helped.

She is now convinced she should never drink anything but "a beer with 55 calories in it."

Ritana2, like many people on this forum, myself included, is more than a bit confused—and also quite forgetful:

Why can I drink one day with no problem? And then I have a reaction the next day, drinking the exact same thing, even from the same bottle! It just seems very weird. I keep meaning to mention it to my doctor, but then I forget. You also get very forgetful after you turn 40! ;) I've written it down now, so that I will remember to ask at my next doctor's appointment. I swear I will.

I think I know what her doctor will say.

NEXT TO THE photocopier in the staff room, Ken is collating short stories for his students while I am printing up papers on the

science of alcohol. We both teach writing here at the University
of Toronto, and I consider him a friend.

"Hey, it just occurred to me," he says. "When your book comes
out, you're going to be *that guy*. Have you even realized that?"

I choose not to answer. Since my sort-of diagnosis of a rare
flushing syndrome, I've become quieter, and somewhat surly—
at least until the first drink of the day, which is now quite late at
night. And anyway, Ken (who teaches a course he created called
Generating Stories from Life) never needs any prodding.

"Dudes in a bar, or wherever, they'll say, '*That guy there*—he
wrote the book on hangovers.' And it'll be literal. It'll be *literal*,
man! Have you even realized that?"

Ken, bless his hard-pumping heart, is one of the few writers
I know who is also a competitive athlete. He has the body of a
rower, the anxiety of an author and "a vested interest in the book
being *done* already!"

So, yeah, I get it. But writing for a living is a tricky endeavor
at the best of times. Then imagine that, for some stupid reason,
you've made your beat the dodgiest slices of the trickiest life
you can find—shantytowns, booze cans, gambling dens, drug
houses—so there's always a good reason to be a bit messed up.
So you're always a bit messed up, until one day you find yourself
genetically modified and managing a nightclub.

"Hey," says Ken. "You still managing that club?"

"Yeah," I say. "Going there right after class."

"Man you're hard-core."

The original idea of the club was threefold: to pay down some
debt, do some research and . . . well, have my own nightclub. But
being a writer, teacher, club manager and single father to a three-
year-old is starting to wear. Half the week, I get to bed at 6 a.m.,
and the other half, that's when I have to wake up. So, no matter

how I cut it, I miss a night of sleep. Laura is starting to become less patient, while the mother of my child will barely talk to me.

I mostly write about hangovers at night, often in the back room of the club, and read about them during the day, when I'm too hungover to write.

Today, I was looking through hangover research papers. There are very few of them, and they all seem to start with a disclaimer about how few of them there are. For example, this report, titled "Development and Initial Validation of the Hangover Symptoms Scale" says:

> Despite its ubiquity, hangover has received remarkably little systematic attention in alcohol research. This may be due in part to the lack of a standard measure of hangover symptoms that cleanly taps the physiologic and subjective effects commonly experienced the morning after drinking. In the present study, we developed and evaluated a new scale, the Hangover Symptoms Scale (HSS), to potentially fill this void.

The scale was created by studying 1,230 students who drink, and inventorying a year's worth of their hangover experiences and symptoms. In so doing, the researchers came up with "a reasonably valid set of adjectives describing common hangover effects." I think they actually meant *nouns*, since the list includes "thirst, tiredness, headache, difficulty concentrating, nauseousness, weakness, sensitivity to light and sound, sweating, trouble sleeping, vomiting, anxiousness, trembling and depression." The average participant reported five out of the thirteen in the past twelve months. I've had all thirteen in the past twelve days. And "red face and welts" isn't even on the list.

The HSS was developed to help study hangover occurrence retrospectively, while the Alcohol Hangover Severity Scale was designed to measure a hangover's potency while its effects are ongoing. In this case, they measure twelve symptoms: "fatigue, clumsiness, dizziness, apathy, sweating, shivering, confusion, stomach pain, nausea, concentration problems, heart-pounding and thirst."

Though I only learned about it today, the scale is quite easy to get the hang of. You just rate yourself on the severity of each symptom, from one to ten, and the total is your score. Right now, talking to Ken at the photocopier, for example, I have an Alcohol Hangover Severity score of 46 out of a possible 120. But again, the scale doesn't take into consideration that, if I were to have a drink right now, my head would start to burn and my blotchy arms would itch and sting.

I can see the dudes in Ken's hypothetical bar: "That guy there—he wrote the book on hangovers." And they're pointing not to the swarthy, barrel-chested, Irish Viking Jew I always thought I was, but the red-faced light-beer drinker shaking in the corner. *That guy there.*

"Hard-core, man," sighs Ken again, adjusting the stapler.

Something's got to change.

"LET MODERATION REIGN!"

It was the poet Horace, amid so much Roman debauchery, who first issued that awkward rallying cry—purposefully ironic in tone, yet desperately sincere in purpose. At the time, it fell on drunken, deafened ears—whereas these days, it seems, the virtue of moderation is the only thing anyone can agree on.

The taboos of the new world aren't so much about the things you do or don't do, but the balance with which you do them. Moderation, above all else, is the new paragon of health, and the logic is apparently impossible to argue with. So, you're into crystal meth/bestiality/Arbonne parties? To each their own, as long as it's done in moderation!

We have also become weirdly aware that what will supposedly kill us one day will be great for us the next, and vice versa. Surely there's been some recent study suggesting that moderate consumption of Pop Rocks will actually decrease your chance of getting Alzheimer's, tinnitus and depression. And of course, this whole ridiculous anti-continuum started with alcohol, it being so naturally, mysteriously dichotomous, and also infused with prescience.

After thousands of years as the most common ingredient in medicines and therapies around the world, alcohol was stricken from the US pharmacopoeia in 1920 as part of the immoderate and dangerous experiment known as Prohibition. Then, almost as abruptly, studies began to appear showing the wondrous benefits of booze.

The list of maladies for which regular, yet moderate wine consumption is recommended grows longer every day. Today, for example, it includes cardiovascular disease, colds, cataracts, macular degeneration, rheumatoid arthritis, Alzheimer's, dementia, diabetes, fatty liver, ischemic stroke, as well as some good old ones like indigestion, insomnia and aging. According to all sorts of science-y sites on the internet, drinking tequila—moderately, of course—helps fight or prevent osteoporosis, type 2 diabetes, dementia and obesity, and is also both probiotic and prebiotic. And beer, according to recent studies, is good for your kidneys, cholesterol levels and bone density, and drinking it is an excellent way to get all sorts of necessary vitamins.

Though medical practitioners of yore did warn that alcohol, no matter how beneficial, be taken "in moderation," that surely meant something very different than the infrequent thimblefuls recommended by today's almost manically cautious boards of health. But there have also been those who argued the benefits of total drunkenness—and even the healthfulness of a hangover.

Arnold of Villanova, writing in the fourteenth century, suggested there was something to be said for complete intoxication, "in as much as the results which usually follow do certainly purge the body of noxious humors." And the Muslim physician Avicenna put forward the same hypothesis, albeit at arm's length: "Some persons claim that it is an advantage to become intoxicated . . . for, they say, it allays the animal passions, inclines to repose, provokes the urine and the sweat, and gets rid of effete matters."

Or perhaps more apt—at least for our purpose—is the wisdom of Oscar Wilde: "Everything in moderation! Including moderation."

THE HAIR THAT WAGS THE DOG
(FA LA LA LA LA, LA LA LA LA)

I'm in Vancouver to visit my folks for the holidays, and I have been making a concerted effort not to drink too much—even if the color it turns my face is appropriately festive. I do have to keep writing this book, though, so I'm carefully selecting my drinking sessions to get the most research out of each one.

Christmas, for example: I've got it all figured out—thematically, at least. Supposedly, chimney sweeps in Victorian London made extra cash selling charcoal to holiday revelers, who would ingest it before drinking. The recipe I've found calls for two tablespoons of fireplace charcoal dissolved in a cup of warm milk.

My nineteen-year-old nephew David has offered to join me on this one. (The drinking age in British Columbia is nineteen, so we may be doing something rather stupid, but not illegal). Also, I keep coming across references to the pre-boozing prophylactic qualities of coating the stomach with olive oil, so I'll be doing that first, then stirring the charcoal into a cup of warm eggnog, while David will follow the standard no-oil method, dissolving the charcoal in a cup of 2 percent milk.

Before any other relatives arrive, I stick my head in the fireplace and scrape a bowl of charcoal. It is supposed to sop up the toxins and poisons—this is why charcoal tablets can help in moments of overdose. But the stuff from the fireplace doesn't dissolve quite as easily as you'd hope, and it turns the milk a dubious purple.

I take my shot of olive oil. It is very oily. Then I chase it with the mug of eggnog and soot. It's a bit hard to get down, but actually doesn't taste too bad. David's teeth have turned black, and he says mine have too, so we gargle a glass of champagne until they sparkle again.

I finish the bottle of champagne while making an anti-oxidental Christmas salad of my own devising. It consists of kale, spinach, chard, pomegranate seeds, walnuts, avocado, mandarin oranges and fresh figs. I lay it out like a Christmas wreath, pour balsamic fig dressing all over it, then get down to the serious drinking, keeping track of David's intake too.

Just in case our preventative methods are lacking in some way, I've been working on a more biblical remedy for the morning: the Gold, Frankincense and Myrrh Cure. Frankincense is a natural anti-inflammatory. It comes in capsule form under the name *Boswellia*. Myrrh is easily available as an essential oil. It increases your white blood-cell count and aids the liver in bile secretion. Then there's the gold: According to George Bishop, "The union of precious gold with fire-inspired distilled wine was considered by the seventeenth-century Spagyricists to be especially effective as a healing agent." Perhaps those wise men were even wiser than we knew.

This is all my idea, but I've elicited a bit of help. I've acquired the frankincense and myrrh—along with some information about them—from Bronwyn, who is a herbalist at a shop called Gaia Garden. She is very smart and patient and not the least bit flaky. And when I said, "Hey, what about gold?" she quickly provided me with the email address of a local alchemist.

The alchemist, herbalist and Ayurvedic practitioner Todd Caldecott, however, doesn't suffer fools quite so gladly. In response to my email query, he wrote, "Okay, but why gold, frankincense, and myrrh, specifically? Is this a Christian book on hangovers?" He then continued: "Gold is used in medicine as a remittive drug, to suppress the immune system. . . . It has a number of nasty side effects and is not recommended for hangovers. Ask King Midas."

But then, in a strange sort of transition, the referential Caldecott seemed to suggest that gold might, in fact, be just the thing: "In Ayurveda, a preparation of gold called *swarna bhasma* is used to decrease inflammation and enhance vitality. In this respect it might be indicated, but because this preparation also contains other metals, current Western medical thinking would prohibit its use."

Of course, one doesn't go to an Ayurvedic alchemist for current Western medical thinking, but rather for a bit of admonishment coupled with a Lord of the Rings reference, which Caldecott finally provided: "In truth, Ayurveda and traditional systems of medicine don't have much sympathy for our vanities, such as thinking we can habitually abuse alcohol to the point of hangover. While liver cells are wizards of regeneration, you can only try to kill Gandalf so many times before you finally succeed."

Whatever the case, and perhaps unsurprisingly, I have been unable to get my hands on any Ayurvedic or even biblical gold so close to Christmas. For this year, at least, olive oil, charcoal, anti-oxidental salad, frankincense and myrrh are all I have to bring. Pa-rum-pa-pum-pum.

But what some people seem to forget is that Christmas is just as much a Viking holiday as it is a Christian one—a purposeful amalgamation of the feast of Yule with the Christ's birthday—and that both were historically debaucherous affairs. In Viking celebration of the Christian wine god's birth, I shelve my new-found moderation and move purposefully into my cups.

David tries to keep up, but he's just a kid. By midnight, I have logged six large glasses of wine, five flutes of champagne, three rum and eggnogs, two whiskies, two brandy beans and probably a couple of turtle doves.

I WAKE THE next morning like a partridge stapled to a pear tree. All I can do is hang there, coming up with stupid lines about Boxing Day.

They call it Boxing Day because you wake up feeling like you've gone thirteen rounds with Mike Tyson.

They call it Boxing Day because last night's bottle is gone, and all that's left is the box it came in.

They call it Boxing Day because that's where you belong: in a goddamn pine box, six feet under.

This is a big one: at least 100 out of 120 on the Alcohol Hangover Severity Scale. When I can finally move, I take the gifts of the Wise Men, but there is no miracle. I lie in my Boxing Day misery, shaking and vomitious, bereft of glad tidings or goodwill toward men. And so, what of the frankincense and myrrh, the figs and pomegranates, the charcoal and olive oil? Well, often, and especially in the very dodgy realm of hangover research, you are your own worst enemy and don't even know it. It won't be until days later, in fact, while digging through my notebooks, that I come across this, from the apostle Kingsley:

There is a great deal of folklore about taking some olive oil or milk before joining the party. This will indeed retard absorption of alcohol, but, as before, it will all get to you in the end. . . . An acquaintance of mine, led astray by quantitative thinking, once started the evening with a tumbler of olive oil, following this up with a dozen or so whiskies. After a couple of hours of nibbling at the film of mucilage supposedly lining his stomach, all the alcohol finally broke through in a body and laid him on the floor . . . I would be chary of this tactic.

The sun has gone down on Boxing Day when I finally leave my bed. I am feeling downright chary—shaky, anxious and nauseous, like a sorry acquaintance of Amis's, or some biblical sucker who misunderstood the Scripture. I drive, very carefully, to the only place that is open, in search of a hail Mary.

I park in front of London Drugs. The doors slide back, and I shuffle into the fluorescence, down the echoing aisles, to the back of this mammoth mega-pharmacy. And there, between specialty orthotics and the morning-after pill, are the bottles of Zwack.

A mysterious emulsion of forty herbs in 40 percent alcohol, Zwack exists somewhere on the herbal booze continuum between Jägermeister and Underberg, which Amis the Elder describes like this: "The effect on one's insides after a few seconds is rather like that of throwing a cricket-ball into an empty bath, and the resulting mild convulsions and cries of shock are well worth witnessing. But thereafter a comforting glow supervenes, and very often a marked turn for the better."

Zwack is like Underberg's slightly more alluring Hungarian cousin: a similar mixture of bitter herbs and alcohol, but with the soft touch of caramel added. It is excellent for soothing a broiling gut, producing that comforting glow with a less violent approach. And in my experience, it is the most effective hair of the dog for squelching gut rot and rejuvenating appetite, all the while curbing that dreaded alcohol down-swoop. And did I mention you can get it at the pharmacy?

Considering the puritanical alcohol mores and bylaws of my hometown, I have no idea how this is possible and don't really want to ask. It may have something to do with the label, which is essentially the same as the Red Cross's red cross, but with a shimmery gold sheen to it. Together with the bright green bottle—more reminiscent now of a large tree ornament—the whole thing looks downright Christmassy.

Also, above the cross is another word, one that somehow rings both crasser and more medicinal: *Unicum*. It doesn't exactly slide right off the tongue, but that's what I ask for at the pharmacy. It makes it sound more pharmaceutical, curative, as though I have no interest at all in getting *zwacked*. But once outside, I revert back to the perfect onomatopoeia for such a satisfying kick.

I take a long, careful glug and feel the heat as the spirit settles inside me. There are, after all, worse things than going a bit red in the face.

YOUR FIRST DRINK OF THE DAY

It is likely—in fact, probable—that one Frank M. Paulsen of Wayne State University has cataloged more hairs of the dog than anyone else in history. In 1960, Paulsen went in search of the elusive hangover cure—or, at least, the myths surrounding it. His research was exhaustive, yet inspiringly rough-and-tumble. He traveled to dive bars, roadhouses and nightclubs in Detroit, Cleveland, Montpelier, Buffalo, Utica, Omaha, Los

Angeles, Quebec, Montreal and Toronto. And he asked for people's advice.

Many of his subjects remained anonymous, since, in his words: "I was unable (or thought it unwise as the case might be) to ask for names and biographical information from over half my informants. The subject matter of the cures and the places where they were collected should justify this high degree of anonymity."

The resulting study, "A Hair of the Dog and Some Other Hangover Cures from Popular Tradition," was published the following year in the *Journal of American Folklore*, and nowhere else since. It is eighteen pages of poetical brilliance: as if Charles Bukowski and Tom Waits sat down to write a story or a song together, then got very, very sidetracked.

Paulsen has a way with people. You can see them opening up to him with a laugh, and you can smell the gin on their breath. And he seems to have hit on a sort of apex in the history of morning-after drinks. Here are a mere five of the 261 hairs he found on his big dog trip—along with descriptions of his interviewees, which, a half century later, thanks to Paulsen's poetic precision, are just as telling as what they had to say:

Cut a big piece of watermelon, if you can get hold of one, and punch holes in the meat with a fork. Pour half a pint of gin over that and eat it down. Be careful of the seeds; they'll kill ya.
—Mr. Meschner, Hawkins Bar, Detroit. Retired; white, male, about 70; born and raised in Detroit, where he worked for the Ford Motor Company for some 50 years; now spends most of his time wandering from bar to bar in the northwestern section of Detroit; quite active for a man his age.

Eat a white Bermuda onion like an apple. Wait half an hour, then take a good shot.
—George Gust, Club 58, Cleveland. Restaurant and bar owner, white, male, about 28; of Greek extraction.

A shot and a beer—a stale beer if you've got any. Break a couple eggs and put them in the beer. Don't eat the eggs. Let [the hangover] fight itself out. But if you get hungry after a couple of doses, eat the eggs.
—Anonymous, Webb Wood Inn. Occupation unknown; white, male, about 65; apparently quite a heavy drinker.

Take cold, jellied consommé and mix in some Worcestershire, celery salt, garlic powder, and about four ounces of vodka. Your secretary will hate you, but you'll be able to get through till lunch.
—Anonymous, Club 58, Detroit. General Motors executive; white, male, about 45.

Mix cinnamon in wine and sip on it. Any kind of sweet wine will do.
—George Fonte, Club 58, Buffalo. Bartender; white, male, about 45; born and reared in Wisconsin; has been tending bar, mostly at exclusive private clubs for 20 years; he was my most cooperative and informative informant. (Fonte also suggested: milk; lemon sherbet; mashed strawberries and sugar dressed with egg whites, gin and chartreuse; a whisky sour; salty dogs; an orange blossom; a shot of vodka mixed with equal parts tomato juice and clam juice; sherry mixed with an egg yolk and served at room temperature; a shot of Pernod mixed with an egg white and four dashes of bitters; and warm seltzer water mixed with bitters.)

It is worth noting that Mr. Fonte really knows his hangover stuff—even earmarking a rendition of the Bloody Caesar, which, to this day, you can usually find only in Canada. But this last one is my absolute favorite (even though the hair of the dog is just slipped in at the very end). You know when people ask, "Who, in history, would you like to have dinner with?" This one is my guy—either him or Lauren Bacall:

You've asked the right man. I sell the cause; I got the cure. When you wake up that way, first go to the store. I mean go to the store, Dad. You open that icebox door, get talking to the lettuce, and you could get mixed up with the milk bottles. That's not good. Buy an avocado, not a hard one, not a soft one. It's got to be just right—tender to the touch, but not too easy. When you get it home, peel it, but peel it gently, so there's plenty of green left. Cut it in slivers, if your hand's steady enough. Salt it light, ever so light. It's better to use your right hand if you're right-handed and you haven't cut yourself yet. Eat that, like it is. Never chill it. It's hard enough to taste it as it is. No, I'm not kidding, eat the avocado. Then you have sex— you know, nothing in bad taste, not fast, not vulgar. Take your time. Lie there a little while; catch a little shut-eye if you can. By no means go again. After you're back down on the ground, I mean really down and relaxed, get up and step into the shower. The water's got to be just right— not too hot, not too cold, just right. Spend a half hour there if you got to. When you step out, walk up to the mirror. Shave. Then use that sexy lotion. You walk out feeling new. You're ready for your first drink of the day.

—John Leon, St. Paul Hotel, Los Angeles, California (by correspondence). Bar manager; white, male, about 35; born and reared in Los Angeles; of Mexican extraction; has been tending bar for thirteen years.

THE HAIR THAT WAGS THE DOG
(ON THE LAST NIGHT OF THE YEAR)

I've never much liked New Year's Eve. It's like St. Patrick's Day, but even worse—pushing inebriated amateurs into a desperately sentimental state of both expectation and reflection. So ringing it in with my lifelong best buddy and my gorgeous, graceful girlfriend at my childhood home seems like the right call—even though my folks are out of town at my sister's for the night. Laura and I would be there with them if it weren't for the Polar Bear Swim tomorrow.

One of the oldest in the world, it is also one of the largest. And if we're going to do it, we've got to do it with Wasko, whom I've known since I was eight. His first name is Mike, but that doesn't matter. He is *Wasko*. And despite him being blond, blue-eyed and articulate, people have always thought we're brothers. This is the 986th time we've had drinks together . . . or something like that.

Mike's actual brother, Nick, lives in an apartment building downtown. It is one of the oldest in the city, and also the closest to where the Polar Bear Swim happens. Hairs of the dog at his place, both before and after the plunge, are a Vancouver New Year's tradition. For Laura, this is all new, and I figure if the three of us are going to ride out next year's first hangover together, we should acquire them together too. But there's just one problem: nobody is getting drunk.

For Laura, this is normal. She'll nurse a glass of pinot grigio until it should be weaned. And Wasko, I'm starting to realize, has curbed his drinking habits since becoming a father, turning forty and taking a job as a small-ferry-boat captain (which he insists is the worst gig to have with a hangover). Now, it seems, he only drinks beer, and much more slowly than he used to. So no one but me has dented the champagne, though it isn't doing the same for me—thanks to the stupid water thing.

There's a rather tired school of thought that goes like this: if you want to drink responsibly and avoid hangovers, have a glass of water for every glass of booze. It's one of those things I've always said, "Yeah, yeah" to, then only pretended to do—and apparently for good reason: drinking water the same way you drink real drinks just sucks. But I've pledged to do it tonight—for my health, sanity and book research.

Of course, the benefit of drinking water is a relatively modern idea. In *Drink,* Iain Gately's indispensable, all-encompassing social history of alcohol, he writes of the ancients, "Water drinkers were believed not only to lack passion but also to exude a noxious odor. Hegesander the Delphian noted that when the two infamous water drinkers Anchimolus and Moschus went to the public baths everyone else got out."

And "everyone else" did have a point. For thousands of years, there was no greater carrier of disease, pestilence and death than water. It poisoned, deformed and drowned you. And now, of course, water drinking will kill you again, the internet suddenly full of warnings that one too many glasses of the stuff can shut down your kidneys.

But more to the point of my current situation are the insights of Dr. Nephew Freud, who counsels wisely: "Attempts to time your consumption to avoid hangovers will lead to severe depression. . . . The advice to 'take more water with it' . . . has a higher irritant potential than any phrase except 'I told you so.'"

"Ugh," I sputter, working through another water glass. But Wasko and Laura are discussing politics, seemingly unconcerned with the ongoing effects of my stupid experiment I can't concentrate on what they're saying. I'm fuzzy, yet sober, lethargic and needing to pee. This one-to-one water-to-booze ratio may be good for you, but it seems to defeat the purpose of drinking, which is to enjoy oneself—or at least feel better. I feel fidgety, bloated and randomly preoccupied. I go for a pee, then Wasko lights a joint.

Even though I grew up in British Columbia, I could never handle B.C. bud, which is famously potent; more than a toke and I'd be trying to take my pants off over my head and jump out the nearest window—like in one of those public-service scare films from the '50s, ending in a puddle of inept paranoia. It just didn't work for me. But over the past few years, living in Ontario, something has changed. I can smoke pot and be okay. In fact, a bit before bed even seems to temper my hangovers. Right now, I'm feeling flat—all too straight and all too heavy—so I go ahead and take a toke, and then another . . .

Wasko (who is a pro when it comes to pot) and Laura (who won't touch the stuff) have switched from politics to art, and I am trying to find my feet. When finally I do, I can't tell if I'm sitting or standing. It feels like I am hovering, in the corner of the room at first, then right between my dear companions, their words bouncing off my head. I try to say something, then have no idea if I've said it or not. I don't even know what *it* is. I try to ask them, "What did I say?" but don't even know if I've managed to say that. I can't read their mouths or the looks on their faces. I watch, but they seem unfazed. I decide that I'm still sitting, seemingly attentive, and haven't said a word. My heart is pounding. I'm sweating and my hands are shaking. I'm sober and wired, wound up and strung out. And then the New Year's countdown starts, the TV suddenly on and blaring: "Ten . . . nine . . . eight . . ."

And now, frantically, I'm trying to reflect on my year. It looks like this: drinking, pitching stories about drinking, writing, being hungover, pitching stories about being hungover, being in the club, trying not to drink, writing about trying not to drink, still being hungover . . .

"Seven . . . six . . . five . . ."

And now I see the year to come: a sober-yet-strung-out string of days. And it's tightening on me, blotched and bloated as I bargain with myself, doctors, editors . . .

"Four . . . three . . . two . . ."

I slap myself. It makes a cracking sound at the very moment the disco ball drops—"ONE!!"—so that Wasko and Laura don't even notice. As they lift their arms in the air, I rush to pop another bottle of champagne. It is the one thing I can do in any state. And as I twist off the foil, this is what I realize: with my general physical state, the self-prescribed drinking, this stupid, pedantic water idea, and now the way-too-powerful-pot paranoia, I have, on this last night of the year, put myself into a state of hangover without even getting drunk—the exact opposite of what I did in Vegas, and nearly as stupid. I am the worst hangover researcher ever.

The champagne pops and bubbles over. Laura laughs. *Fuck the water thing!* I think, glugging down the fizzing booze. I give her a kiss, then pull Wasko in for a hug.

"Sorry," I blurt.

"For what?" he says.

"For trying to drink responsibly. It won't ever happen again."

"*That*," says Laura, "is your New Year's resolution?"

A REASONABLE GO UNDER THE CIRCUMSTANCES

If there is one thing I've learned from my research this year (and there really may be only one thing), it is how difficult it is to plan even the most basic hangover. By the very nature of drunkenness, controlled experiments lose their controls very quickly. You can have everything all set—and then the set, drink by drink, starts to fall apart, often in dramatic fashion. There is a reason we use the Spanish word for "flask"—*fiasco*—the way we do. And that's something that even the most attentive booze researchers have yet to get a handle on: correcting for chaos.

In fact, until just the past decade, the actual scientists were making a lot of the same mistakes that I am. Out of the very few laboratory studies on hangovers, practically none of them made sure their subjects had reached a blood alcohol level (BAL) of zero—so it's very likely that some of their subjects were in the same state of "hangover" as I was while driving a race car—i.e., drunk. You would think at least a booze scientist should understand the distinction between drunk and hungover. But strangely, no. And even if they had, laboratory testing has proved to have other, less avoidable problems.

One is that when people know they're being tested for hangover research, and the study involves them attempting to complete some task—which is usually how you get results of hangover on cognition—people don't apply themselves the way they normally would. In the real world, people try to override, supersede and disguise their hangovers to get things done. When they know they're being tested for hangover, however, they tend to *let* themselves be hungover, and barely even work to overcome it.

Another big problem with testing hangovers in labs is what's generally referred to as "research ethics–imposed consumption levels," which basically means that most university or publically funded researchers can't get you drunk on much more than a six-pack of beer. This mitigates the chaos, sure, but hardly allows for the spectrum of hangover possibilities. And also, people don't get drunk-then-hungover in a laboratory the same way they do at home, in a hotel room or at their buddy's place.

In all these senses, naturalistic studies like the ones I've been devising make more practical sense. Except for the *fiasco* aspect of things. Most science-based naturalistic studies rely on subjects recalling their consumption and activity from the night before. This presents a multitude of obvious chaos-infused, selective-memory, straight-out-lying hindrances. Even *I* can't be sure I'm not subjecting myself to them.

Dr. Richard Stephens, senior lecturer at Keele University in Manchester, England, and his hangover research colleagues have a few recommendations for this. One has to do with biomarkers—things found on your body that indicate what you've been getting up to. One, called ethyl glucuronide (EtG), can be located in your skin and hair and then tell a person in a lab coat precisely how much you had to drink last night. The testing was devised for things like parole and mandated rehab, but its use would make naturalistic hangover studies that much more reliable.

Another recommendation, and one that Stephen's team has been using itself, is to not let your subject know that you know that they have a hangover. As such, they've scheduled their cognitive testing of student volunteers, who do it for a small fee, for Thursday and Saturday mornings. The students have no idea that the tests have anything to do with hang-overs. But the testers know that the traditional binge-drinking nights at this particular university happen to be Wednesdays and Fridays.

As Dr. Stephens wrote in a recent paper, "Pilot work in our laboratory shows that a significant percentage of participants recruited in this way arrive at the laboratory with a hangover, but as they are blind to the study aim of assessing hangover effects, they are as close as possible to how hungover people behave in the real world, aiming to have a reasonable go under the circumstances."

THE HAIR THAT WAGS THE DOG (AND THE SORRY, SCATTERED POLAR BEAR)

So now it's the new year—the morning of the Polar Bear Swim— and I am standing in my parents' kitchen in front of what looks like the detritus of a hipster farmers' market: little bundles of kale, berries, ginger, ginseng, organic eggs and plenty of protein, magnesium and vitamin powders—the year's refined accumulation of morning-after ingredients for my ultimate Shaky Shake. It used to include a couple shots of booze, but that was before my apparent gene mutation.

The Shaky Shake is something I've been working on for quite a while now—my own ever-changing concoction of antioxidizing, antitoxifying, hopefully hangover-solving ingredients. And today, it turns out I need it. Despite last night's water-method discomfort and circumstantial pot paranoia resulting in pseudo-hangover symptoms before intoxication, I still managed to get right drunk, and then actually hungover.

On the other side of the kitchen counter stands Laura before an even stranger assortment of ingredients, a sheet of paper in her hands. "Um . . ." she says, looking back and forth, from the paper to the counter: elk meat, ostrich meat, ground chicken, tripe, minced vegetables and powdered vitamins.

My family takes in strays—beasts of all appetites and quandaries. Over the years, I've returned home to a potbellied pig, a wolf and a traveling musician or two. Right now, most of the animals are at my sister's with the rest of the family, and only the cats are here with us. But they are special cats.

Tonka (a Maine coon who weighs more than most dogs) cannot be fed before Yoda, who has to be fed in the bathroom. Trousers has to be fed outside, but can't stay outside after that. (My parents spend a lot of time standing on the back deck, yelling "Trousers!" to the confusion of anyone new to the neighborhood.) Arrow will hide, then try to steal everybody else's food, and should be fed in the master bedroom. Oscar is a gentle psychopath with twenty-two toes, who will eat and swallow almost anything and, for everybody's safety, has to be kept, most of the time, in a very large cage. They all eat different combinations of raw exotic meats and supplements. And should you mess up any part of it, bad things will happen.

Laura, trying to be helpful, offered to feed the cats. So my mother wrote instructions on that piece of paper, which Laura is looking at as if it might burst into flames. "Um . . ." she starts again.

"Um, what?" I say. Or maybe I snap it. Strewn about me on the floor are pots and pans, pieces of Tupperware and a dozen dog and cat bowls, but not a blender. "Where the hell is the blender?"

"I don't know!" says Laura. "I'm trying to feed the cats!"

Then, as if conjured, the lunatic felines surround us. And now it's a *Katzenjammer*, the wailing of cats and, I believe, the root of the German word for "hangover," so that even this one—the first of the year, the one I stringently prepped for—has already become ironic, iconic and downright Wagnerian. I'm sweating profusely as I send a text to my folks: "Happy New Year! Everything is fine. And also, where is the blender?"

My sister's cabin has notoriously impossible cell-phone reception. A half hour later, a text comes through: "Happy New Year's to you! The blender broke five years ago. Sorry. We can certainly get a new one. We love you! Is Trousers indoors?"

By then we're in a maelstrom of fed-up felines and dubious directives. I give up on my Shaky Shake, do my best to help with the cats, re-embrace the concept of the hair of the dog—Drunkzilla be damned—and pack a bag of booze for the Polar Bear Swim. Laura, a little scratched-up and headachy, is nonetheless ready to go.

"Let's do this thing," she says.

QUITE A FEW years ago now, I was put in charge of organizing Wasko's bachelor party weekend. It became a distinctly gonzo affair. A dozen or so of us packed up cars with various intoxicants and firearms and drove to a cabin in the woods north of Vancouver.

The first night started early in the day and went quite hard for a very long time. This was before I started researching hangovers in any serious way, but I'd paid close attention to one aspect of the terrain. The next day, after dragging ourselves out of various bushes and woodland burrows, we started the trek to a location I'd marked on a map. Maps of these parts of Canada are not like maps in the rest of the world, and a good deal of dead reckoning is still required. Hungover as hell, we prodded each other forward and downward through the woods, our feet skidding over cliff edges, until finally we heard it: the thundering churn of a glacial waterfall.

In Canada, there are places where, because of topography and gravity and the magic of the world, water that should be frozen remains somehow liquid, cascading off a cliff for thousands of years. And where it lands, it burrows, creating glacial pools so deep you could never reach the bottom—at least, not without freezing to death. If you want to jump into liquid ice, this is where you do it. And if there were ever a reason, it would be to cure a bachelor party hangover. This, at least, was my thinking at the time.

But before we get to the waterfall, and down the last slope, into the pool at the bottom, let me tell you what I've learned since then. In the first-ever episode of National Geographic's recent TV show *You Can't Lick Your Elbow*, two healthy young men performed a rigorous series of exercises that included sprinting, lifting and as many push-ups as they could do. Then they were given a two-minute break, wherein one of them was told to run into the sub-zero ocean. On a second go of the same exercises, the one who froze himself in the ocean was able to keep on with the push-ups far longer than the one who didn't. There was more to it than that—I'm repeating this from memory—but the results and the reasoning were as such: frigid temperatures activate what we like to call the fight-or-flight mechanism, flooding the body with adrenaline.

Dr. Herman Heise and his colleagues, writing in the *Journal of the American Medical Association*, concluded that when confronted with an emergency situation, the intoxicated individual can indeed address a more coherent, seemingly sober front to the problem at hand. "The fast sobering period," they explain, "is caused by the abundant secretion of epinephrine from the suprarenal glands located above the kidneys. It temporarily counteracts the effect of alcohol but not the percentage in the body fluid."

Epinephrine is adrenaline. And it appears, at least to me, that this "fast sobering period" caused by the fight-or-flight mechanism is just as relevant, if not more so, in regards to hangover. Dr. Stephens, in Manchester, agrees that my theory makes sense: "If you shock the body, the heart beats faster and the bodily system speeds up the rate of blood flow. That'll carry all that glucose you're so in need of, so lacking when hungover, to everywhere you need it. That makes sense." And when taken to adrenaline extremes (such as jumping in liquid ice, or off a very tall building), could it create a complete physiological reboot? "It sounds kind of plausible," says Stephens. "But I don't know how you'd get evidence of that . . ."

On Wasko's bachelor-party morning after, we wound and stumbled our way down a tree-lined cliff, along a hundred-foot waterfall, into the gulley, until we reached the rushing bottom. And then we leapt, with various degrees of trepidation, into the whirling water hole. I will never forget the face of Mike's brother, Nick—a tattooed, Spartan badass—as he clawed and screamed to get back out. His eyes were those of a gerbil spinning on an electrified wheel. For me, the cold was overwhelming and painfully invigorating. All ten of us jumped in and burned in the freezing water as we screamed.

Then we got out and started the long hike back. And on the way, we were laughing, calm but rejuvenated, loose and floating, so that the journey up was so much easier than the one down. And we were ready for the day.

That, I guess, is the point of the Polar Bear Swim.

THERE ARE A couple hundred of us in bathing suits, or approximations thereof. To our right is a guy in a tuxedo, then some Mexicans wearing Viking helmets; to our left, a group of skinny blondes in sombreros. People are randomly singing football songs and "We Are the Champions," and the vibe is generally jovial. You get the sense that most of us are still more drunk than hungover. And yet, everyone's waiting. There's no rope holding us back, but we're waiting for the whistle. It's kind of like at the start of the running of the bulls, except we're polar bears, and the ocean is as cold as the Spanish sun is not—okay, it's not at all like the running of the bulls, but everyone is waiting for the whistle.

And then, finally, it sounds, and we run, screaming and howling into the waves. Laura and I and the Waskos and our other
friends are at the front of the pack, and we get out far enough
with pure forward thrust before the cold hits our innards. I dive
under and come back up, and hug Laura, who is trying to catch
her breath. She stays out there with me, surrounded by the roaring bears, while Nick scrambles back for shore, then Mike after
him. Especially remembering the glacier pool, it is not too cold—
just cold enough, in fact, to numb the flesh and blow the New
Year cobwebs out. When Laura's shivering becomes practically
erotic, we stride back to the shore and join the boys.

Wasko has lost one of his sandals. He's good-natured about it,
but I've known him and those sandals long enough to see he's a
little upset. So, while the others are grabbing for towels, I'm scan-
ning the surface of the water for a well-designed hiking sandal.
I spot a plastic lei, a beer can, what looks like a plastic Snoopy
doll . . . and then, whoomp, there it is, bobbing on the waves like
the severed foot of a sunken sailor.

"Hmmmm," says Wasko, but not smoothly like that, because
his teeth are chattering, and because his sandal is a ways offshore.
"Maybe I'll just leave it."

There are very few things I excel at. And since starting the
research for this book, that list seems shorter every day. But diving back into difficulty will always be one of them. So I run and
dive, and stay under the whole way.

When I hand Wasko his footwear, he gives me a good old
polar bear hug. I wish I'd brought my sandals from Toronto too,
because we're heading back to the building now, and pulling
socks and shoes over wet, sandy feet is annoying—so I decide to
forgo it and just walk barefoot through the streets.

"You're going to regret that," says Nick, the badass worrier,
gesturing to my feet.

"This man *deals* in regret," bellows my best friend, Wasko, in his stage-practiced baritone, slapping me on the back as we leave the beach. It's a good line, and everyone's laughing. But it's almost too precise, like one day I'll have to put it on my business card, right beneath *Writer* and *Hangover Researcher*: *Regret Dealer*. I squeeze Laura's hand as we cross the street, and I make my first New Year's resolution: I'm going to quit the club—managing it, that is. I'll tell her later. The decision is made. And it feels healthier than any hangover shake.

We go back to Nick's apartment building and have some hairs of the dog. The first I've had in a while. I choose an Irish whiskey, just because I feel like it: "The dog," as Kingsley Amis put it, "is of no particular breed: there is no obligation to go for the same drink as the one you were mainly punishing the night before."

I couldn't agree more, and particularly with the "mainly" part. I hate when people ask, "What's your drink?" It's like asking what your music is.

I follow the year's first whiskey with a cleansing beer, and still the flushing symptoms are nowhere in sight. No redness, nor itchiness. Not even a blotch. Maybe the Pacific pushed it right out of me.

"How are you feeling?" says Laura.

"Cured," I say. And pour us a couple mimosas.

A NEW YEAR'S REVELATION

Over the past year, I've been trying out all sorts of remedies—patches, capsules and elixirs marketed specifically to cure hangovers. Some seem to lessen some symptoms, but it's always hard to tell, and I haven't quite figured out my research protocol yet.

As such, the first draft of my New Year's resolutions includes trying to mitigate the *fiasco* factor in future experiments, accumulate as many products as I can, isolate the active ingredients in these products, learn about them and test their efficacy. This being somewhat wordy, I've distilled it down to: *Do more, and do it better.*

Laura has already returned to Toronto, and I'll be heading there tomorrow. But first, I've decided to use the opportunity of a last dinner with my parents to start on my resolution. In front of me are what I have acquired so far: packages of dried prickly pear, ginkgo biloba leaf, milk thistle, kudzu root and Japanese raisin seeds. These, along with something called N-acetylcysteine are—as far as I've been able to find—the most common active ingredients in hangover remedies from around the world. To properly ingest them, I'd need to create powders, teas, emulsions, tonics—except for the N-acetylcysteine (NAC), which comes in nice little capsules. So I've decided to start with that.

According to some nurse-y friends of mine, NAC is an amino acid supplement often used in the ER to treat drug overdoses, and a couple websites I've found say it mops up those free radicals left after the chemical warfare of a night of heavy drinking, but it should be taken at the beginning. The nice guy at the health food store wasn't sure about all this, but said he'd send an email to his supplement rep and get back to me. I've also read that combining NAC with Vitamins B_1, B_6 and B_{12} improves its efficacy. My mum only has B_6 and B_3 in her cabinet, so I take those along with the NAC as appetizers and wash them down with a sip of wine.

But now, after just two minutes and two more sips of wine, something is starting to happen—something very familiar, but much, *much* worse. It is like my head is on fire—or more like it's stuck in a goldfish bowl of boiling tomato juice. Through the red-glassy haze, I can see my mother mouthing something. Something like "Oh my God! Are you alright?" But I am not alright. I am a flushed goldfish, a red mutant. Drunkzilla has returned—and this time it's personal.

I get to the sink and soak my head under the cold tap for a while. Then I take some antihistamines and sit back down at the table. My parents are looking at me in horror. "*Roar-r-r-r*," I say.

THE NEXT DAY, I receive a forwarded message from the nice guy at the health food store:

> Vaso, NAC upregulates glutathione production, and glutathione helps the body deal with acetaldehyde which is the primary cause of hangovers. As you can imagine, curing hangovers isn't exactly a high priority for the academic research community, so these theoretical links haven't exactly been the focus of clinical trials.
>
> Also, to be entirely frank, I'm pretty sure that the author should be researching this stuff himself if he wants to write a book about it. George

There is no mention of any side effects. But between George's snark and this brand-new case of Boiling Tomato–Asian Fishbowl Syndrome, I decide to strike N-acetylcysteine treatment off my list, at least for now.

But then, later that day, while contemplating Marmite, a revelation of sorts occurs. Marmite and hangovers have a lot in common. They're both mysterious, misunderstood and a by-product of booze. And those who never acquire the taste might say they're both bloody awful.

If you've never tried Marmite, do so carefully. It has a tar-like consistency and a strong, unique flavor: bitter and salty with something else—something intense and inexplicable, like the Earth's core, or dark magic. Most people outside the UK have never tried Marmite, or else tried it once and immediately condemned its existence—tasted the devil within and despised it with every inch of heart and palate.

But my mother being British and very involved in my upbringing, I've always known that Marmite is a virtuous thing. Made from yeast extract (a by-product of beer brewing), it is the most concentrated edible source of nutrients you can get. And long before I started researching hangovers in earnest, it was my go-to morning-after breakfast. Bereft of a blender on New Year's Day, I'd whipped up some Marmite and toast instead. And now, as I do so again, I consider its obscure, dark-corner place as a sort of next-level, edible hair of the dog. After all, Marmite comes precisely from the thing that bit you—but with far more vitamins and minerals, and not a drop of alcohol. I study the familiar label, examining more closely its magic ingredients.

Marmite contains a yeast extract, sodium chloride, vegetable extract, spice extracts, celery extracts, folic acid, vitamin B_1 (thiamine), vitamin B_2 (riboflavin), vitamin B_3 (niacin), Vitamin B_{12}—wait, *what*?

This jumble of letters, heretofore meaningless, suddenly pops into acute relevance: *Vitamin B_3 (niacin)*.

My mind begins to reel, the film flipping backward from the pills I took last night to the months of Shaky Shakes to the Asian allergenist . . . allergist—*whatever*! Marmite has 6.4 milligrams of niacin in it. I run downstairs and locate the big canister of vitamin and mineral supplements I'd been using for my Shaky Shakes before coming to Vancouver, where there is no blender: 20 milligrams of niacin. Then I go to my mum's bottle of vitamin B_3. It is a 50-milligram dose. That weekend trip with Laura—when the first symptoms hit—was also when I started using powerful B multivitamins.

Fucking *niacin*!

The 6.4 milligrams of the stuff found in Marmite appears to be just fine, and probably helpful. But much more than that, it seems, and one could start to mutate—or at least exhibit some uncomfortable symptoms. So, finally, I have my equation: *A bit too much booze + way too much niacin = scaly, itchy, fiery Drunkzilla.*

I am happy to have solved this riddle. But I am also full of *Verschlimmbessern*—a German neologism that combines *Verbessern* (to make something better) with *Verschlimmern* (to make something worse) so that the new meaning is "to make something worse by trying to make it better." It means coating your stomach with olive oil. It means making morning-after smoothies with high doses of niacin. And as far as I can tell, it means everything when it comes to trying to cure a hangover. Overreaching will fill you with *Verschlimmbessern*. And it's hard to come back from that.

I make a Marmite sandwich, wash it down with a glass of Zwack and start to work on some new New Year's resolutions.

AND UP SHE RISES

What shall we do with the drunken sailor,
What shall we do with the drunken sailor,
What shall we do with the drunken sailor,
Earl'eye in the morning?

It is an age-old question. But still the answer remains at bay—if not miles offshore. Even in a somewhat sober sing-along, most people hit the second verse and become instantly incoherent:

Put him in the . . . thing . . . with the other thing on it . . .
Put him in the . . . oh wait—
Put him in bed with the captain's daughter!

That's pretty much all anyone can remember; the most recognizable sea shanty ever, and nobody knows the words. Plus, it being the morning, our sailor is probably hungover rather than drunken. Of course, back when intoxicated English sailors were still worth a shanty or two, the scurvy bastards didn't yet have a word for this distinction. But like frat guys of yesteryear, they had a lot of stupid ideas:

Put him in scuppers with a hosepipe on him . . .
Put him in the bilge and make him drink it . . .
Shave his chest with a rusty razor . . .
Heave him by the leg in a running bowline . . .

Though most of these things only ever existed in the middle of the ocean, one can assume they're pretty nasty to have done to you at the best of times, let alone with a hangover. Even that somewhat salacious one isn't what it seems; it turns out the captain's daughter is a euphemism for lashing someone with a cat-o'-nine tails.

Of course, there's a history of this kind of brutality on dry (but not dry enough for some) land too. Down through the ages, we've spent a lot of energy coming up with torments for those who are already tormented. In the earl'eye morning of the Ottoman Empire, they poured molten lead down a drunkard's throat. Charlemagne had drunks whipped for their first offense, and repeat offenders were pilloried. In 1552, public drunkenness became an offense under British law, punishable by stocks rather than penance. Thereafter, the town stocks were used mainly as a public drunk tank, so the offender could sober up in that perfect medieval state of increasing shame and pain.

This was corporal punishment as spectacle—meant to serve as a warning to others—and the banging, hanging head of a man in the stocks could very well be the root of the word *hang-over*. That said, there are plenty of other possibilities too.

In 1555, an exiled Swedish bishop in Rome named Olaus Magnus the Goth published a very long, very weird book with a ninety-nine-word title, the first six of which are *A Description of the Northern Peoples*. Chapter Thirty-nine, Book Thirteen, entitled "On the Punishment of Drunkards," contains the following admonishment:

"There is no good cause for anyone to become inebriated," writes Olaus. "From frequently imbibing too much wine men's breath stinks; the French become impudent, Germans quarrelsome, Gotar mutinous, and Finns lacrimose. Perhaps every drunken man who is detected in these various failings deserves to suffer suitable punishment for his foul and filthy behaviour."

And the suitable punishment? The inebriate would be "placed on a seat with raised points and be lifted up by ropes, holding a huge drinking-horn, full of beer, in his hands, given to him as he sat there, so that he might drink it off very quickly or otherwise stay sitting uncomfortably on that very sharp seat." Another possible root (albeit obscure) of the word *hangover*: the act of being forced to down a hair of the dog–filled horn while literally hung over a seat of spikes.

Then there was the drunkard's cloak—a wine barrel turned into a straitjacket, as described in an American Civil War–era newspaper: "One wretched delinquent was gratuitously framed in oak, his head being thrust through a hole cut in one end of a barrel, the other end of which had been removed; and the poor fellow 'loafed' about in the most disconsolate manner, looking for all the world like a half-hatched chicken."

In 1680, Massachusetts Bay governor John Winthrop referred to one Robert Cole, who, "having been oft punished for drunkenness, was now ordered to wear a red D about his neck for a year." Some scholars point to this humiliating necklace as the true origin of *The Scarlet Letter*. And then, too, this letter hanging around the neck of a drunkard could also be the basis of being *hungover*.

Despite all these rather traumatic possibilities, English-speaking academics have only been able to date the use of *hangover* as "the after-effect of drinking too much" to 1904, first appearing in an obscure book of humor called *The Foolish Dictionary*.

THE FOOLISH DICTIONARY.

BRAIN The top-floor apartment in the Human Block, known as the Cranium, and kept by the Sarah Sisters — Sarah Brum and Sarah Belum, assisted by Medulla Oblongata. All three are nervous, but are always confined to their cells. The Brain is done in gray and white, and furnished with light and heat, hot or cold water, (if desired), with regular connections to the outside world by way of the Spinal Circuit. Usually occupied by the Intellect Bros., — Thoughts and Ideas — as an Intelligence Office, but sometimes sub-let to Jag, Hang-Over & Co.

There are very few other references until the start of World War II. And then, up she rose. So much so that if you punch *hangover* into a Google doohickey called Ngram Viewer, which charts the history of word usage in books, this is what you get:

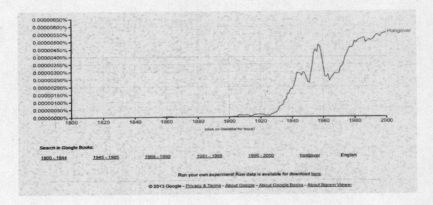

Few other nouns have this kind of mountain range. And it's an odd one to traverse. But although it took all this time to find it, it's hard to deny how well the word works—due in part to its intrinsic, even accidental, double meaning—the idea of being something left over from before, a sort of detritus; the image that of your flotsam-filled head just hanging there, from the stocks or a wine barrel, or over the edge of a toilet bowl.

James Harbeck, author of the blog *Sesquiotica*, puts it much more sesquiotically:

> *Well,* hang *surely has the right kind of tone—it rings, it smacks of execution and hangnails, it has echoes of* harangue *and* dang, clang, bang, *et cetera. And* over *carries tastes of finality as well as of impending and threat. The word, when said, is strong on the first syllable and then gives a weak double-beat to follow up. It's not really a dactyl, though, a three-time beat; it's more*

like a half note and two quarter notes. It's a bit like the sound of a hammer being let drop: a big bang and then two little bounces.

Probably what being swung onto a spiky chair sounds like, or falling over in a wooden barrel, or being heaved by the leg in a running bowline, earl'eye in the morning.

PART FOUR

A MAD HATTER IN MIDDLE EARTH

IN WHICH OUR MAN IN YE OLDE ENGLAND FORGES STEEL,
THATCHES A ROOF AND FALLS DOWN SOME RABBIT HOLES.
APPEARANCES BY A BLACKSMITH, A DRUID
AND BILIOUS THE OH GOD.

The English are a hedonistic, freebooting, swashbuckling race,
who, when not allowed to go around the world conquering
other people, drown our sorrows any way we can.
—JULIE BURCHILL

MY HANDS ARE SHAKING AND MY HEAD IS ACHING as I swing the hammer down. It rings off the anvil—*a big bang and two little bounces*—glancing off the end of my stock. "In again," bellows the blacksmith, his long hair and muttonchops the same fiery red as the sparks raining down around us. Despite his name, Richard Wood is a man of steel and fire—a philosophical blacksmith who heads the fine arts department at Plymouth University, a self-made metallurgist who also gives lectures on computer 3-D printing.

"It's not about getting it right," he says. "If you're not failing, you're not trying hard enough. The value of failure is incalculable." In which case, I am incalculably well off. I plunge my steel back into the forge.

The forge, at its hottest spot, is two and a half thousand degrees, and now the steel is glowing starlight-bright. Sweating, I pull it out and try to steady myself, to quiet the ringing in my head. This is not a good way to treat a hangover. And today I didn't even intend to have one.

I am in Devon, England, firstly on a magazine assignment, to research and write a story on ancient hand skills, and secondly to test out the concept that people in olden times didn't suffer modern-day hangovers because of what and how they drank. Richard Wood is happy to back this up. "As long as I'm working, I never feel hungover," he says. "I just sweat it out in the forge!"

As part of this experiment, I've put aside the product testing for the moment—left all those little packages of ingredients on the other side of the ocean—and will be using only what I find here in the Old World. I even got a hold of the right stuff to drink. Two days ago, in London, I met with a mead maker. The plan was to drink it through the day while being an ancient blacksmith, work it out in the forge, get to bed early and wake up without a hangover.

But then yesterday, things went sideways, as they sometimes do. This time, it was in a Victorian-manor-turned-decadent-Wonderland, with Lewis Carroll–themed suites and rooms—each with its own free, fully stocked minibar. *Free. Fully stocked.*

As Adam Rogers puts it in his introduction to *Proof*: "It's not enough to admire the pretty bottles filled with varicolored liquids. . . . You're supposed to ask questions about them—what they are and why they are different, and how people make them. The only people who can get away with going that far down a rabbit hole are journalists, scientists, and three-year-olds. And three-year-olds aren't allowed in bars."

As the journalist in the Jabberwocky Room, I had my own questions to ask those bottles. But I'm still not sure of the answers. And now, after a long, hungover drive through the Devon countryside, I am here in Middle Earth, banging away at steel. I have explained very little of this to the blacksmith, who probably just thinks I'm uniquely inept and shaky.

Though he teaches in the town of Plymouth, Wood lives in a tiny, hidden hamlet that feels very Hobbittish. The long, downward-winding tunnel of trees through which we drove was so thick my car lost all radio signals—nothing but wind, birds and the babbling of a brook. The ancient sign above the door

of his cobblestone dwelling reads, THE CIDER HOUSE. Across
the way is the Brooksleigh Manor, for which all things named
Brooksleigh—be they towns, villages, hamlets or manors, anywhere in the world—are apparently named. Ye Olde England
doesn't get much older than this.

In the spirit of this place, I tried to hunt down some witches,
and also a nearby druid or two. Professor Ronald Hutton, a
member of the Council of British Druid Orders, teaches in Bristol,
not far from here. So I sent him a message asking, "What might
a druid do for a hangover?"

He wrote back promptly:

Dear Mr. Bishop-Stall,

Thank you for your question, but, alas, I know nothing about this
subject (perhaps fortunately!).

With every good wish,

Ronald Hutton

Even closer to here, in Boscastle—famous for being hit, not so
long ago, by a devastating flood—is the world's largest museum
of witchcraft. In quick response to my query, I received this:

Dear Shaughnessy,

Nothing comes to mind regarding hangovers—there are charms
related to drinking, such as putting St. John's Wort in wine to alleviate
melancholy—but I will discuss this with our Assistant Curator and get
back to you. We are happy for you to visit the Museum although a
tour might be a little tricky as we are a bit low on staff tomorrow. We
will see what happens when/if you get here . . .

Best wishes,

Peter

Could it be that the dark arts are becoming, at least in England, too politically correct to face the force of hangovers? That thought, along with the slightly unnerving "when/if" followed by ellipses, set me instead on the trail of a bad witch—or at least one who writes a blog called *A Bad Witch's Blog: A Blog about Paganism, Witchcraft and the Day-to-Day Experience of a Witch Living in the UK.*

The bad witch also responded promptly. But for various reasons, I shall not divulge what she said just yet.

WHEN I PHONED my son to tell him what I'd be doing today, he asked me to make him a sword, but according to Wood, that would take about thirty hours of labor. We've decided on a fancy-handled fireplace poker instead, and I'm still working on the pointy end. But one of the nice things about blacksmithing, and why it might not be too bad with just a low-grade hangover, is how very forgiving it is. Sure, if you grab the wrong end, your skin won't forgive you, but the metal always does.

"It isn't what people think," says Wood. "They talk about steel like it's an immutable part of the world—linear, unaffected—when really its strength is in its malleability. It allows you to go back and try again."

And so, as the ringing subsides, I breathe, embrace the heat and give it another shot. Then, with a few good, steady swings, I finally get the point.

FEUDAL EFFORTS

We tend to look down on the Dark Ages. For one, we call them the *Dark Ages*, evoking hundreds of years of ignorance and misery. But they might also be seen as a golden era in the history of hangovers: when most of Europe aligned religion, nature and science to create systems of drunkenness that worked *for*, rather than against, the drinker. Much of this had to do with the Crusades, as alcohol became a meaningful divider between Christianity and Islam. Jesus was a god of wine, after all, and the enemy was sober.

By 1200, Benedictine monks had become the world's largest, most established wine producers, creating organic appellations of remarkable quality for every nobleman in England, Europe and beyond. For those in the upper class, drunkenness was embraced as a noble duty: a responsibility, by birthright, to revel in divinely bestowed decadence that others could barely imagine. Lords became drunk as lords—coining a phrase to last a thousand years.

And since beer, ale, mead and cider were so clearly better than water at sustaining healthy laborers, the commoners got better booze as well. Bylaws were passed to make it more affordable, and royal tasters roamed the land to test the quality of the drink. It was more common then to find an ale seller in the town stocks for providing crummy hooch than one of his customers for drinking too much of it.

None of that is to say that the drinkers of the Middle Ages were immune to the morning after. They took hangover treatments from ancient times and made them a part of their every day—particularly cabbage, herring and cheese. But perhaps their wisest practice was to have two sleeps.

In 2005, historian Roger Ekirch published *At Day's Close: Night in Times Past*, which reveals that for most of human history, in almost every culture, we slept in two distinct chunks of time: a first sleep shortly after dusk, a period of wakefulness, and then a second sleep until the sunrise. And apparently, people did all sorts of things between these two sleeps—they read, painted, played music, had sex, ate and, most certainly, mitigated a hangover.

The effect of alcohol withdrawing from one's system and interrupting deep rest would probably have coincided with the waking period between the first and second sleep. Rather than waking up anxious and fighting to get back to sleep, our ancestors could have embraced those hours, as well as a lover, perhaps a hair of the dog, and then gone back down for a second, rejuvenating sleep, so that the morning after was actually the second time waking, and certainly less of a shock.

Wrote Sydney Smith, a nineteenth-century cleric: "I start up at two o'clock in the morning, after my first sleep, in an agony of terror, and feel all the weight of life upon my soul. . . . But stop, thou child of sorrow and humble imitator of Job, and tell me on what you dined. Was there not soup and salmon, and then a plate of beef, and then duck, blanc-mange, cream cheese, diluted with beer, claret, champagne, hock, tea, coffee and noyeau? And after all this, you talk of the mind and the evils of life? These kind of cases do not need meditation, but magnesia."

And also, no doubt, a nice second sleep.

A MAD HATTER IN MIDDLE EARTH
(AT A BOOZY TEA PARTY)

After a night drinking with the blacksmith, I am somewhat hungover. But I am also fairly well rested, sipping from a glass of Bacchus wine, eating cheese off a barrel and looking out on fields of grapevines and cows grazing from here down the valley to the River Dart. It is a bright blue day and the sun is high in the sky.

On my way to thatch roofs, I've stopped at Sharpham Vineyard and Cheese Dairy for the first day of English Wine Week—despite whatever oxymoronic jokes such a thing might elicit.

According to the blacksmith Wood, Sharpham is an organic winery, and the Bacchus grapes I'm drinking come from vines just behind his Cider House. They are not, in fact, fully organic. Not yet, but Sharpham is hoping to get there, and doing as much as they can the way it was done in the Middle Ages, with a few notable idiosyncrasies: instead of *Benedictine* monks, Sharpham now has Buddhist ones working the vineyards. They handle the grapes zenfully while the milkmaids milk the cows.

After the Bacchus and cheese, it's time for high tea, complete with cheddar scones and sparkling white. On the patio behind me, a four-piece swing band is playing—stand-up bass, guitar, clarinet and trombone—and when the old man in the bright blue suit starts to sing, his voice is like herbs, champagne and sunshine. This, right here, is how to treat a hangover.

WINE AND CHEESE

They've been produced and served together, very purposefully, since we've known how to make either. In ancient Greece and Rome, grated cheese was added right to the wine as a preventative against its ill effects, both in the drinking and the morning after. This custom continued well into the Middle Ages, and then morphed eventually into the ubiquitous wine and cheese party, wherein we scarf down cubes of Gouda and glasses of merlot, unaware of the reasoning behind it, let alone whether such reasoning is sound.

But Tim Spector knows. The British genetic epidemiologist sees the hidden sense in all of this. But he's also getting ready to leave his office at King's College for a well-deserved vacation in southern Spain, so I promise to keep it brief.

"Can you just tell me about wine and cheese?" I say, by way of Skype.

"Well," says Spector, "it starts, as things often do, at the microbial level. We are made of microbes, and our guts are full of them. What we've found is that, when we drink alcohol, certain microbe species stimulate the immune system as if it were under attack. This results in inflammation, and toxins get leaked out into the intestines. Also, certain microbes replicate when we drink alcohol, particularly one called *Erysipelotrichia*, which happens to produce dehydrogenase—the enzyme that breaks down alcohol to acetaldehyde—which as you probably know is the catalyst for a hangover. And studies have also shown that when mice are fed these toxins, they seek out more alcohol than normal mice, suggesting that these particular microbes, which make us feel so terrible when we drink alcohol, might actually be encouraging us to drink more alcohol . . ."

It is probably worth mentioning that geneticists have a different understanding of *brief* than anyone else on the planet, except perhaps geologists. Spector really drives this home when I ask the seemingly obvious question: What the hell kind of purpose could such sadistic microbes possible serve, genetically speaking?

"Well," says Spector, "you've got to remember what a very new phenomenon alcohol is for the human body."

Having just started to come to terms with how very ancient drinking and drunkenness are, how much a part of our shared humanity, this briefly baffles me. But of course, ten or twelve thousand years is the blink of an eye to a geneticist. "We are made of microbes, and for over a million years our immune system has communicated through the ones in our guts, warning of dangers. And now we've got this relatively new situation, it seems, where the breakdown of alcohol is triggering these warnings and starting a whole system that causes symptoms which make us feel very sick."

Spector believes these reactions are getting progressively worse—that our microbial balance is out of whack and our gut enzymes are failing because of what we've done to them. "Whether it's hangovers or dust allergies, our immune responses are going a bit haywire—I think because we've just been hitting them with way too much, including all these antibiotics."

"And what about the cheese?" I ask.

"Well, just look at the mice," says Spector, and for a moment I think he's joking. But he's actually referring to studies on mouse guts and alcohol and probiotics, and how these things might affect hangovers.

"But wait just a minute," I say. "How can you tell if a mouse is hungover?"

It is the perfect setup for a dozen good punch lines, but Spector refuses to even nibble at the bait. "We can't," he says. "So we look at liver damage."

Since a mouse can't tell us when it feels hungover, but we know that hangovers and liver damage are systemically similar—insofar as both are caused not by alcohol, but by the chemical products created when it is broken down—researchers like Spector take the leap and use signs of early liver damage as a marker for hangovers. "Experiments making mice binge-drink like humans produced signs of liver damage," says Spector. "Again because of the toxins, leaky gut and inflammation."

But here's the thing: when the researchers fed those same binge-drinking, microbe-filled mice "probiotics containing beneficial microbes such as lactobacillus" (science-speak for well-aged cheese) along with all that booze, the mice showed virtually no harmful effects whatsoever.

It is this kind of thing that suggests maybe nature's got our backs after all—and all a man or mouse really needs is a bit of cheese with his wine.

A MAD HATTER IN MIDDLE EARTH (HELL-BENT FOR LEATHER)

After a day of thatching houses, then drinking a batch of Thatchers Cider with Charles Chalcraft, the master thatcher who thatched the pub in which we sipped our ciders, I have now moved from the realm of boozy British tongue twisters to that of gothic lateral-thinking puzzles.

This place, in fact, feels like a horror film set designed by M.C. Escher: impossible ancient walkways connecting rooms full of dangling hooks, vats of animal fat, piles of skin and cartilage and hair, great bubbling pits of tar-like tannins and green acidic goop in which a soul could disappear forever. I try very hard to fall into none of them. And then there's the smell: this omnipresent potpourri of cow flesh, lime, burnt hair and oak bark. It is the only thing more distracting than my own imagination. I would not want to do this job with a hangover.

"No, you wouldn't," says Andrew Parr, the owner of the J. & F.J. Baker tannery. "And we start at 7 a.m." Also, skin—even a cow's—doesn't have the forgiveness of steel. "Once it turns from hide to leather, it can never change back. That is the alchemy of tanning."

This is one of the oldest working leather factories in the world. Abandoned—along with mead halls and vineyards—when the Roman Empire fell, it functions now much as it did back then. I am interested in all of this for the article I'm working on, but also because of tannins and the part they play in the mysteries of booze and hangovers.

Oak tannins, which here turn animal skins into perfect leather, are also what turn clear, fairly tasteless alcohol into a glowing, flavorful whisky—while grape tannins, which create a robust and flavorful red wine, are now being blamed (along with sulfides) for causing red-wine hangovers and triggering migraines.

"Tannins are in the skin of all plants," says Parr—a tall, thin man with a scientific mind and a humble, yet oddly poetic delivery. "The leaves of a flower, the skin of a grape, the bark of an oak. They protect plants with enzymes and antibodies. They're kind of magic. Nobody really knows why oak tannins bind so well with animal hide and turn it into leather. We don't even know how we know this. One can imagine it, though back when this island was one big forest and Britons were hunters, eventually they'd notice all those strange things happen; an animal pelt left in a puddle by an oak tree would take on certain characteristics—strength and suppleness. That's probably how we learned to make leather."

From drinking just enough last night, today I've got a bit of the Shining—that incandescent, divining-rod feeling. And as Parr talks, and the synapses fire, I start to get glimpses of something: connections of craft, Old World ways of doing things, mice and cheese and organic transformation. I am thinking backward from whisky to wine to mead, and then forward to the possible triggers of modern-day hangovers.

But then, all this thinking, along with the smell of this place, is starting to give me an Old World headache. So, for now, I let it go. I get back in the car, roll out of Middle Earth and head for the big city.

THE HERE AND HEREAFTER

If the physical hangover—due to changes in human biology, environment and alcohol potency—has shifted over the years, the metaphysical hangover has undergone revolutions. And nothing has had a greater impact than the idea of drunkenness being a sin. But where the hell did that even come from? It isn't one of the deadly seven, nor mentioned in the Ten Commandments. And when Dante outlined every circle of damnation and who would end up where, there was no mention whatsoever of drinkers, or even drunkards.

As with so many churchy things, the change most likely came about because of power and money. As the Dark Ages ended and feudalism gave way to new systems of commerce, people were now free to open all sorts of businesses—the most popular being pubs. Suddenly, there was somewhere to get together other than church, and a place to dump coins other than collection plates. And that's when the clergy started shouting "Sin!"

Sermons of the time referred to taverns as "the Devil's church, where his disciples go to serve him." And by the 1600s, the devilish dangers of drunkenness were so oft-spouted that, apparently, they warranted a greatest-hits compilation.

Put together by Samuel Clarke and Samuel Ward, *A Warning-piece to All Drunkards and Health-drinkers Faithfully Collected from the Works of English and Foreign Learned Authors of Good Esteem* contained no fewer than 120 of the best worst things that could possibly happen should you dare to have a drink. In the short run, they included stabbing your mother, being run over by your own horse, and vomiting so violently that veins popped in your head. In the long run, they were much, much worse.

At the time, booze was referred to as "health"—a semi-irony that seems to have driven the Samuels slightly crazy: "You promise your selves Mirth, Pleasure, and Jollity in your Cups," they goad. "But for one drop of your mad Mirth, be sure of Gallons and Tons of Woe, Gall, Wormwood, and bitterness here and hereafter. . . . You pretend you drink Healths, and for Health; but to whom are all kind of Diseases, Infirmities, Deformities, pearled Faces, Palsies, Dropsies, Head-aches, if not to Drunkards?"

And in case you'd dare to bring up those pesky commandments or deadly sins, the Samuels beat you to it, and then beat you over the head with them: "It is no one sin, but all sins, because it is the Inlet and Sluce to all other Sins. . . . God would be pleased to open the Eyes of some Drunkard, to see what a Dunghill and Carrion his Soul is."

One Drunk Vomiting broke a Vein after 2 days great pain Dyed.

4. being Drunk were Slain by Carts.

One Drunk Rideing over plowed-lands fell and broke his neck.

a Child that murthered his Mother he being Drunk.

The clergy, through Christian guilt, had got a hold of a mighty weapon—one that Islam had been wielding all along—and amplified its power: a hangover wouldn't just hurt anymore; it would burn with regret, remorse and the threat of eternal damnation. Your morning-after suffering was instant penance, an example of God's wrath—and, even more profoundly, a little bit of the hell now waiting for you in the afterlife.

A MAD HATTER IN MIDDLE EARTH
(MORE NUTTON THAN NUTTS)

By the time I reach London, it is raining, like it only rains here: that kind of downpour that feels unleashed by magic, to wash the cobblestones of pestilence and ale. The early lunchtime drinkers on the sidewalks hold briefcases over their heads as they push back into the pubs.

In the doorway to the Wellcome Library, I check my phone for messages. The last one from Dr. Nutton came a week ago, and—as if predicting this future—ends with the caveat "If wet, meet me by the desk near the entrance." Wet, I go in and sit by the desk.

While I wait, I keep checking my phone. Just last month, Dr. Nutton made headlines for discovering an ancient Egyptian papyrus that appears to be one of the oldest prescriptions for hangover ever. But even more than Dr. Nutton, I'm hoping to hear from Dr. Nutt. Apparently, this is the kind of slant-rhyme silliness you find yourself in, should you seek out hangovers in jolly old England.

Dr. David Nutt has made headlines too. Lots of them. As drug and alcohol czar to the British government, he did so by saying that alcohol is more dangerous than psychotropic drugs, and by pronouncing ecstasy to be safer than horseback riding; more headlines for being sacked by the British government; a few more for suggesting that terminally ill people should take LSD.

And then, over the past couple years, the headlines have looked like this one from the *Daily Telegraph*: "Get Drunk

Without the Hangover on Professor Nutt's Pill." The accompanying articles suggest that Nutt has created a synthetic form of alcohol that can get you tipsy without any ill effects, and that he has also created its antidote. So one pill makes you drunk, and one pill makes you sober, and neither gives you a hangover. If there's anything to it, this could be the future of drinking, the Holy Grail of the morning after. Or he could just be a nut.

The man behind these headlines, though, is utterly elusive; it's like trying to make a date with Willy Wonka. For over a year now, I've been sending him messages. At times, he'll respond, just a word or two, and then he slips away. He knows I'm finally in London and hoping to meet with him, but now he's disappeared again. Down a rabbit hole, through a door of perception? Who knows? This is England, after all; every doorway is a magical wardrobe, a phantom tollbooth, a looking glass—on one side a Nutt, on the other side Nutton . . .

"Hello!" says Dr. Nutton, shaking off his brolly. "So sorry I'm late!"

FROM HIS ECCENTRIC teeth to his bespectacled blue eyes, Dr. Vivian Nutton is classically British in that odd, Old Worldly way. He rolls his Rs at the beginnings of words, and also in the middle. "These were found," he says, turning the pages of the book he's brought, "rrroughly 120 years ago, in Oxyrrrhynchus."

Oxyrhynchus is the site of the first systematic archaeological dig in Egypt. The treasures discovered include a massive garbage pile, from which archaeologists have gathered hundreds of boxes of thrown-out papyri that now sit in some Oxford basement that I imagine to hold the same air as that warehouse at the end of *Raiders of the Lost Ark*. They remain untouched until people like Nutton decide to open them. His resulting research is "the larrrgest single collection of medical papyri ever to be published." And in amongst the pages is an ancient prescription for hangovers.

"Here it is," he says, turning to a photo of a piece of papyrus. "As you can see, it is a column of something bigger—the sides have been lost—with fifteen lines of Grrreek in it. And it was written by somebody with quite a neat hand." I peer at it. Dating to around two hundred years before Cleopatra, it is all ancient Greek to me, but not to Nutton. "The actual word for 'hangover' is rrreprrresented only by the last letter," he says. "But the length of the line and the fact that we're dealing with something to do with the head suggests that the word in front of it was *hangover*."

"They had a word for 'hangover'?" I say.

"Well, not prrrecisely," he acknowledges. "More a suggestion. It would translate as 'drrrunken headache.'" I ask him to translate the whole, as best he can. "Rrright," he says, and puts his finger on the page: "*For a drrrunken headache: wear leaves of Alexandrrria chamaedaphne—which is a sweet-smelling shrub— strrrung together.*"

I've heard this before, from the disciples of Pliny—not this wording, exactly, but the concept: that the ancients believed certain plants and herbs warded off the ill effects of wine. This coincided with the idea that intoxication was caused primarily by noxious fumes rising to the brain. According to Clement Freud, "The poet Horace when inviting Virgil to dine at his house, tells him not to forget his anti-fume remedy." Garlands and laurels around the neck and body, and also wreaths around the head, were worn for the same purpose.

"The notion," says Nutton, "was that the scent from these plants would counteract the ones that made you sick. Smell was seen as something physical, the active substance itself. But there may also be an aspect of magic to it. It is very hard to draw the line between folklore and what might be descrrribed as scientific."

This, of course, is true in many cultures, and the entire history of treating hangovers. According to Freud, the pre-Socratic philosopher Democritus not only formulated an atomic theory of the universe, but also posited that hangovers could be cured by giving the essence of *sarmentis* (a kind of twig) to the sufferer, as long as it is done "without his knowledge."

Andy Toper, meanwhile, writes that Indigenous North Americans dealt with the morning after by "placing grated horseradish on the forehead, binding it tightly, then placing your thumb in your mouth and pressing it firmly against the roof."

In Puerto Rico, it is supposedly still customary to treat a hangover by squeezing the wedge of a lime in one's armpit. Voodoo, as practiced in Haiti, recommends putting pins in the cork of the bottle that got you. And Robert Boyle, the seventeenth-century scientist, founder of modern chemistry and pioneer of the modern scientific method, provided this (apparently very scientific) treatment for the bottle ache: "Take green hemlock that is tender and put it in your socks, so that it may lie thinly between them and the soles of your feet: shift the herbs once a day."

In further regard to British voodoo, apparently some English witches (inspired by tales from Haiti) have tried to transfer particularly bad hangovers to people they didn't like by hammering nails into effigies. Perhaps more helpful would have been a visit from Bilious the Oh God.

Bilious, born of British fantasy writer Sir Terry Pratchett's noble imagination, is a mournful deity who takes on certain people's hangovers (perhaps thus explaining the 20 percent of humans who, according to studies, are impervious to them). His followers are a hungover lot who moan "Oh God" in hopes that he'll relieve *their* torment as well. The Oh God literally wears his burden—a toga besmirched with the excess food and drink that oozes out of him daily.

I ask Dr. Nutton if he has heard of Bilious, but he has not. So
we return to the book at hand, and Alexandrian chamaedaphne.
I've read about garlands and wreaths made of violet, roses, ivy,
laurels, even cabbage leaves, but this is very specific.

"It is," he agrees. "It is part of what makes the passage special.
I think, because of the way it is written, you could say with some
certainty that this was something that was not only prescribed,
but often used. It still grows in Egypt."

"I'll have to get some," I say.

"But not just any chamaedaphne," clarifies Dr. Nutton.
"You'll want the one from Alexandria."

"Be it for the scent or the magic," I say.

"Prrrecisely."

BY THE TIME I check in to my hotel—of which I'm supposed to
be writing a review—the rain has stopped, but I am soaked right
through. The suite I'm given is luxurious in a deliberately retro,
dark and sexy way. Part 1970s Santa Fe, part colonial Africa, it
is unabashedly kitschy, manly, opulent and dangerous—the kind
of place Joseph Conrad and Ernest Hemingway might have had
drinks together before going into the jungle or out for a night on
the town. That kind of suite.

As I hang my dripping clothes above the giant bathtub, a knock
comes at the door. I find a bathrobe, pull it on and answer. Before
me is a beautiful woman in a black cocktail dress, and inexplicably
she is presiding over an impressive yet compact, genuine cocktail
bar. The walls of the hallway behind her are fashioned out of
heavy brass molds, both concave and convex, depicting medical
implements: surgical tools, vials and beakers; anatomical figures,
busts and skeletons; and also parts of the human body arranged
as if on display—bones, teeth, eyeballs and organs. The overhead
light dances off them in a lovely, haunting way.

Long before it became a private club and hotel, this place was a Victorian hospital and treatment facility specializing in venereal disease. I happen, at the moment, to be reading Robert Louis Stevenson's *Strange Case of Dr. Jekyll and Mr. Hyde*. Although Jekyll's tastes are "rather chemical than anatomical," his house was purchased from a celebrated surgeon, and what was now his laboratory used to be an operating theater. I imagine its hallways much like this.

"Good evening," says the woman in the black dress, smiling at my bathrobe. "Would you like me to come back later?"

"No," I say. "I don't know. What for?"

"It is complimentary cocktail hour. Do you like gin?"

"Yes," I lie.

She gives a push, and the bar glides into the room.

LONDON BURNING

It has always been volatile and dichotomous—a beacon of progress in a cloud of fog and swirling vapors; a labyrinth of mystery, wonder, soot and booze. According to a thirteenth-century tourist, London had "only two curses: fire and drunken idiots." And that was before gin even existed.

The technique of distilling wine down to its essence started with Islamic chemists. Then Franciscan monks used this alchemy to harness what they saw as the fabled fifth essence. It was named, simultaneously, *aqua vitae* (the water of life) and *aqua ardens* (firewater)—a magical, life-giving incendiary four times more powerful than any liquid known to man. And then the world started breathing fire.

Already heavy drinkers, the Brits became soused, wrecked, plastered, pickled, pie-eyed, paralytic, blotto, stinko, four sheets to the wind, fogmatic, gutter-legged, hit by a barn mouse, holding up a lamppost, drunk as lords and just plain smashed.

By 1723, statistics showed that every man, woman and child in London consumed at least a pint of gin per week. By any standards, this suggests mass lunacy. Wrote William Lecky in 1878, "Though small is the place which this fact occupies in English history, it was probably, if we consider all the consequences that have flowed from it, the most momentous of the 18th century." Gin shops announced that you could get drunk for a penny, dead-drunk for two, and have straw for nothing. Beneath them were straw-strewn cellars, into which "those who had become insensible were dragged, and where they remained till they had sufficiently recovered to renew their orgies."

Three different Gin Acts were passed, all of them failing in one way or another to stem the drunken tide. Wrote Barbara Holland, quoting an unnamed "contemporary" of the time: "Being illegal, it was now made not so much from malt as from 'rotten fruit, urine, lime, human ordure, and any other filthiness from whence a fermentation may be raised. . . .' Turpentine was the most popular flavoring. Sulfuric acid was added for its kick. Nobody counted the customers who went blind or dropped dead."

It is safe to say that the gentle medieval hangover, coddled by a regimen of mead and ale quality control, wine with cheese in it, and two guilt-free sleeps a night, had, by the eighteenth century, turned into a tormented, sinful, vein-popping, passed-out, gin-soaked, snake-in-the-boots, shaking, waking nightmare.

In 1751, William Hogarth produced and published a couple of etchings—what Iain Gately calls "Jekyll and Hyde images of drinkers"—that would become iconic in the art and argument of temperance. The first etching is entitled *Beer Street*. It depicts good Englishmen of the Empire out on the avenue, industrious even while having a drink. A nationalistic, hardworking, money-making, fun-loving group, they are going about their jobs while drinking from overflowing steins. In fact, the only person pictured without a beer in his hand is an aloof, skinny painter, who is working on a poster advertising gin.

In the second etching, *Gin Lane*, all hell has broken loose. A cripple beats a blind man. A boy fights a dog for a bone. A mother feeds her child gin. A madman dances with a baby impaled on a spike. A toddler howls as her mother is lifted, skeletal and naked, into a coffin. Bricks are falling from the rooftops and the barber has hanged himself in his attic. The only ones plying their trades are the pawnbrokers, prostitutes, gin sellers and morticians. In the center of it all, a bare-breasted, syphilitic mother lets her baby tumble headfirst into the subterranean stairwell of a gin shop. At her feet dies a skeletal pamphlet salesman. An unsold stack of cautionary leaflets slips from his basket—their title: *The Downfall of Mrs. Gin.*

By 1750, twenty million gallons of gin per year were being consumed in Britain—that's more than twice as many as today, despite a population growth of nearly 600 percent. The scope was almost unfathomable—an adolescent empire unhinged by its own binging. And whereas the denizens of Beer Street got up the next morning a little fuzzy in the head, laughed at their own good-natured sport and got back to work, the zombies and lunatics of Gin Lane were gone for good—deformed, dead and forever debauched, burning in a whole new state of hangover hell.

A MAD HATTER IN MIDDLE EARTH
(AND A COUPLE OF STRANGE CASES)

The beautiful woman in the black dress has packed up her movable bar—a teak chest that unfolds like Inspector Gadget's elaborate drink cart—and left me here with an icy glass of gin, mint, crushed ice and some other stuff. Although I tend to have an aversion to gin, this is a very good drink—like a strong, dusty mojito. I sip it and walk through the bedroom onto a large smoking deck with jade plants and creeping vines, to listen to the sounds of Soho.

Before the hospital was built, this land belonged to John Harrison, the inventor of the marine chronometer, which made it possible to establish a ship's longitude at sea, and thus changed the world. I know of him by reading about the rum trade and pirates, and what should be done with a drunken sailor. Where once was his estate, and then a hospital, is now this place: the Hospital Club—with seven floors, four bars, a restaurant, film studio, screening room and art gallery. It is co-owned by the co-founder of Microsoft and the co-creator of the Eurythmics: a dark, comfortable labyrinth designed to inspire cutting-edge creativity. Apparently, Radiohead has a studio on the top floor.

In looking at some maps, I've found that not only is the Hospital Club right between the fictitious homes of Jekyll and Hyde, but also equidistant between the site of Jack the Ripper's first victim and the most famous address in Britain: 221B Baker Street. In the introduction to my edition of *The Strange Case of Dr. Jekyll and Mr. Hyde*, Robert Mighall describes the streets below as "a foggy, gaslit labyrinth where Mr. Hyde easily metamorphoses into Jack the Ripper, and Sherlock Holmes hails a hansom in pursuit of them both." The air out here is thick, yet cool. It feels like the city is vibrating.

I go back inside to send a few messages—one to my boy, one to my gal and one to the British Gut Project. Started by Dr. Spector (who should be kicking back on a Spanish beach by now), its purpose, in conjunction with the American Gut Project, is to study the bacterial diversity of bellies on both sides of the Atlantic. I want to see if they'll fit my Canadian one into the mix as well.

As Spector put it to me about my gut, "If even part of this hangover quest is personal, you don't want to ignore that bit of your body that is 90 percent of your cells and 99 percent of all your genes." And also, there's that flushing thing. After my New Year's niacin revelation, I thought I had it figured it out. But recently, I've been feeling those symptoms again, just not so extremely, and I couldn't help asking an actual geneticist about it.

Spector politely traded out sudden gene mutation for the possibility of a change in my microbes and their response to alcohol. "To me, that possibility is very likely," he said. "But the only way to really know for sure is to let us study your gut." And so, ghoulish as it may sound, that is the process I am starting. I fill out an online form, send it off to some folks in white coats and decide to mix another drink.

There are two minibars, both in extravagant glass cases. I open the one containing four bottles of premixed elixirs. The other, which can only be opened with "the key to pleasure," contains a velvet blindfold and paddle, a leather riding crop, nipple clamps and some other things I haven't yet identified. The radio is playing Chopin. It's been a long week, yet a quick and blurry one too—as though I was swallowed by a minibar in Devon and tossed out here in Soho; the Jabberwocky Room to the Jekyll Suite.

I turn up the music. This is my only night in London for a while. I make myself a Vesper martini, invented in *Casino Royale* by that charming, well-groomed sociopath on His Majesty's Secret Service. I unpack a dress shirt, find my cuff links (labeled SCOTCH and SODA) and have a drink while I shine my shoes.

WEREWOLVES OF LONDON

The mysterious transformative power of alcohol can change your shiftless uncle into a competitive break dancer, coax uncomfortable truths from the lips of your accountant or reveal the stand-up comic in your shy office colleague.

Such relatively minor metamorphoses are due mostly to alcohol's disinhibiting effect on the cerebral cortex, or as William James—glorying in the glory of getting drunk—puts it, "its power to stimulate the mystical faculties of the human nature, usually crushed to earth by the cold facts and dry criticism of the sober hour. Sobriety diminishes, discriminates and says no; drunkenness expands, unites and says yes. It is in fact the great exciter of the 'Yes' function in man. It brings its votary from the chill periphery of things to the radiant core. It makes him for the moment one with truth."

But what if that truth is a brutal one? And the shift not an upward, barbaric yawp lifting you into the realm of "Yes!" but a twisting transformation into some dangerous beast, with little or no awareness of your other self? Regaining consciousness into a bad morning after—not knowing where you are, how you got there or where those cuts, bruises, broken-off hood ornaments and torn business cards came from—can feel like the waking of a monster.

In *An American Werewolf in London*, the titular tourist wakes up after a full-moon night naked and confused in a wolf's den at the London Zoo. After climbing out, hiding in the bushes and stealing a boy's balloon and then a woman's full-length dress, he finally makes it back to the London flat of his British girlfriend—but neither of them has any idea what happened the night before. They figure he must have been drinking.

Most of us, as the haze dissipates and we start to feel more human, can piece it together on our own. But others are cursed to stagger through the day never knowing what happened. As little as we've learned about drunkenness, we know even less about the switch that flips into blackout and turns your basic hangover into a dark, tangled mystery.

Trying to solve that mystery is the main premise of most werewolf flicks and also hangover ones—including *Remember Last Night?*, *What Happened Last Night?*, *Dude, Where's My Car?* and, of course, *The Hangover*—in which our heroes' blacked-out depravities aren't fully discovered until the credits roll. And then we see what a man in his cups can become.

Samuel Clarke, he of the 120 ways to leave your liver, sermonized about drunkards that alcohol "dehominates them . . . and so disfigures them, that God saith, *Non est hæe Imago mea,* this is not my Image." It is the transformation of human into demon, gentleman into werewolf, Jekyll into Hyde.

The good doctor in Robert Louis Stevenson's tale sees the switch not due to the devil, but instead a chemical carving-out of his own base self—echoing the popular idea that booze doesn't create, but rather reveals and amplifies what already exists, sometimes for good, but often for very bad. Jekyll explains the elixir he's discovered: "The drug had no discriminating action; it was neither diabolical nor divine; it but shook the doors of the prison-house of my disposition."

When Jekyll downs the transformative juice, he becomes like a man intoxicated for the very first time:

> *I felt younger, lighter, happier in body; within I was con-scious of a heady recklessness, a current of disordered sen-sual images running like a mill race in my fancy, a solution of the bonds of obligation, an unknown but not an inno-cent freedom of the soul. I knew myself, at the first breath of this new life, to be more wicked, tenfold more wicked, sold a slave to my original evil; and the thought, in that moment, braced and delighted me like wine.*

And of course, it's not long before Jekyll's alter ego—a stooped, hairy beast freed from the chains of humanity—starts stalking the Soho streets, where this Hospital Club now stands. They are the same rainy sidewalks, here below me, where Warren Zevon—in his only ever Top Forty hit—saw a werewolf clutch-ing a Chinese menu.

If you know Zevon only from "Werewolves of London," with that catchy *a-woooooo* chorus howl, it might be easy to dismiss the lyrics as esoteric silliness. But in the pantheon of truly inge-nious hangover songwriters, Warren Zevon is up—or perhaps down—there with Tom Waits, Nick Cave, Kris Kristofferson, John Prine, Bessie Smith and Kanye West.

Every Zevon line holds meaning—and no less at all for his biggest hit. In the fourth verse of "Werewolves of London," he name-checks Lon Chaney, and then Lon Chaney Jr.—and has them walking alongside the Queen.

Chaney Sr. was the first real monster of Hollywood—on the screen and possibly off it too. He was known as "The Man of a Thousand Faces." His only son was not born Lon Chaney Jr.—and nearly not born at all. Most stories have it that the infant

Creighton Chaney was stillborn and came to life only when his father rushed out to a partially frozen lake and submerged him beneath the surface. Others have it that he was trying to kill the baby—or at least that this is what Cleva Creighton thought, which plunged the young mother into depression, madness and alcoholism, finally culminating in her suicide attempt by swallowing mercury while Chaney Sr. was on stage. Then he divorced her, got sole custody and told Creighton she had died.

There is no doubt Chaney Jr. had it tough from the start, but trying to figure out the truth is like looking at a man with a thousand faces. It is at least clear by Junior's own admission that he feared his famously monstrous father, and that it wasn't until after Chaney Sr.'s death at a young age that Junior learned his mother was alive—and finally went into film, which Chaney Sr. had forbidden. The studios wanted to change his name to that of his famous father. Instead, he used his own name for leads, another one for stunts, another for extra work, and yet another for walk-ons. Then, finally, impoverished and exhausted, he relented. As he told a Hollywood reporter, "They had to starve me to make me take my father's name."

And then Lon Chaney Jr. became the next great monster of the movies—the only actor to play every one of the big four: Frankenstein, Dracula, the Mummy and, most famously, Larry Talbot, Hollywood's first doomed lycanthrope, in *The Wolf Man* (1941). That's what made him even bigger than his dad—and spun five *Wolf Man* spin-offs. But along with the sins of the father, a whole host of demons followed—from set to set, monster to monster.

Said Junior's co-star, Evelyn Ankers, "When he wasn't drinking, he was the sweetest. Sometimes he hid it so well." Director Charles Barton witnessed the blackouts: "By late afternoon he didn't know where he was."

In the original *Wolf Man*, Larry Talbot, distressed, asks his doctor if he believes in werewolves. The doctor's response is perhaps evasive for Talbot, but eerily apt for Lon Chaney Jr.: "If a man becomes lost in the fog of his own mind, he might imagine he is anything."

After completing *Abbott and Costello Meet Frankenstein* in 1948 (the last time he would play the Wolf Man in a feature film), Chaney Jr. tried to kill himself—a suicide attempt sparked, as his wife profoundly put it, "by the emotional exhaustion he suffered during his transformation scenes."

When the man finally did die, his body was donated to science. The University of Southern California still keeps his liver in a jar to show what extreme alcohol intake can do. And there is no grave for the rest of him—not the Wolf Man, not Lon Chaney Jr., nor even Creighton Chaney, who'd apparently been dead on the day that he was born.

But of course, like his father, he'll live forever on the screen—and also in that song: both of them on parade through the streets of London, checking their hair in the window of a pub, beasts in human form before the transformation.

PART FIVE

TWELVE PINTS IN TWELVE PUBS

In which our man in postmodern England talks to
a shrink, then drinks a lot of beer while traveling to
the end of the world—or, rather, the World's End.
Appearances by Dr. Richard Stephens, Simon Pegg,
Nick Frost and at least one robot bartender.

GARY KING: *Don't you miss it? The laughs? The camaraderie?
The fights? The hangovers so fierce it feels like your head's full of ants?*
PETER PAGE: *Well, maybe the first two.*
—The World's End

THE ALCOHOL HANGOVER RESEARCH GROUP WAS formed in 2009 when Joris Verster of Amsterdam reached out to ten of the top hangover researchers in the world— or rather, *the* ten hangover researchers in the world. Dr. Richard Stephens, senior lecturer in psychology at Keele University in Manchester, is their man in England.

He has longish, springy brown hair, Polo glasses and a boyish smile. His cell-phone ringtone is a jazzy little number that is the same as that of the main character on *Episodes*—a TV show about a sardonic yet optimistic Brit with springy hair who moves to L.A. to write a TV show. His accent brings to mind— at least to the mind of a shallow North American journalist— John Lennon without the edge. Especially when he's talking about his band.

"As a teenager in Liverpool, I was in a band," he says, leaning back in his office chair. "The rehearsal space we had was in a bingo hall that my friend's mum worked at, so we didn't have to pay for it, but the catch was we could only get in there between nine and twelve on a Saturday morning. So we would roll up in various states of hangover. And what we found—and we all commented on it—was this sort of glowing feeling of being invincible, at least creatively. Do you know what I mean?"

Psychologists tend to be more anecdotal than other scientists, which is helpful to people like me. I tell him I know exactly what he means, and in fact I've already named the phenomenon, on page something-or-other of this very book: the Shining—that kind of low-level hangover that opens up your senses to the creative electricity of the world. Stephens is intrigued by this. It fits nicely among the concepts of *his* upcoming book, *Black Sheep: The Unexpected Benefits of Being Bad*, and also his ideas about executive function.

In psychology, executive function is a hypothesized set of mental controls that essentially allow you to multitask. They help you prioritize by focusing on a certain set of stimuli while inhibiting others. So alcohol, being one of the great *dis*inhibitors of all things human, should theoretically mess with this process.

"There are, of course, tasks where you want to follow a logical line of thought," he says, motioning toward his desktop, "but if you're trying to be creative, the opposite may be true; sometimes you want to link seemingly unrelated concepts and find divergent ways of thinking . . ."

"And maybe," I continue for him, "it doesn't stop there. Perhaps it can last right through the morning after. Ergo, the Shining—and bands from Liverpool." I like talking to this guy; it's not every day I get a spot in the choir.

"That makes sense," says Stephens. "I believe executive function is strongly affected by hangover, but the research hasn't really been done."

What he means is, it hasn't been done well. There *are* studies out there that were meant to test, among other things, this part of cognition. I've read them all: Danish truck drivers with long-haul obstacles; aviation pilots made to drink bourbon and 7UP and then play around with shapes on a screen; merchant marines presented with a lot of Anheuser-Busch products (both alcoholic and non), and then asked to deal with a simulated power-plant problem. (My favorite thing about that particular report comes as a sort of aside, under study protocol: "Participants with a positive BAC level or pregnancy test on arrival were excluded." Imagine being the soldier rejected by a merchant marine hangover study because you're already too drunk—or it being how you discover you're going to have a baby.)

The odd thing is that none of these studies show any detrimental correlation between hangovers and the ability to prioritize and multitask. But that might just be due to their unnecessary weirdness. "There are much better ways to study these things," says Dr. Stephens, who is partial to something called the flanker task, whereby participants are made to focus on a target on a screen while "distracters" are presented both nearer and farther away. Their inability to do so in regards to "nearer compatible distracters" is, to Dr. Stephens, an indicator of executive function disruption.

Partly in the business of busting myths, Stephens's recent study of studies is taking on two of the most popular beliefs when it comes to hangovers: that women get them worse than men, and that they're harsher for old people than young. In both cases, he believes this simply isn't true—or rather, that if it *is* true, it's essentially these two groups' own fault. Based on the same blood-alcohol content, they should all generally feel the same; it's just that women are generally smaller, and some old people tend to drink more—not more than teenagers, but more than most people. This is where he starts to lose me.

"I believe the age thing is a fallacy," he says. "Hangovers are, for the most part, a young person's malady. And drinking follows a horseshoe distribution." This does not mean it's a lucky one, but rather U-shaped—with heavy binge drinking in the teenage and college years followed by a long dip upon embarking on all those busy and upstanding things like jobs and kids, and then a slow rise as you get sick of your job and your kids, until finally the kids move out and you retire and get back to the business of drinking. Unless, of course, your job is writing or rock 'n' roll, wherein, I'd assume, a roller-coaster or even flat-line distribution would be more apt.

"It's important to realize [and here the doctor loses a bit more of me] that people get wiser as they get older. They learn from things. So although they may consume the same amount over the long run, teenagers binge and old people drink steadily through-out the week—so their BAC doesn't spike in the same way. I have learned what I can drink, and how much, and so I don't really get hangovers anymore."

"Good for you," I say. "But why do so many people my age go on about how their hangovers are getting worse?"

"I think they just don't remember. Like with childbirth, they've buried the memory of the pain. But that's just a guess. It's very hard to find funders for that kind of research. Really, you'd need to get old people and young'uns into a study. My suspicion is there'd be no difference."

But I, having recently turned forty, am worried Dr. Stephens may be wrong.

IF YOU PUNCH "drinking over 40" into a search engine, the first four headlines (at least today) are subtitled "Why Hangovers Hit Harder"; "Why Hangovers Hurt More"; "It Gets Worse"; and "Why Hangovers Are Such a Bitch as You Get Older." Dr. David W. Oslin, professor of psychiatry at the University of Pennsylvania, summarizes some recent find-ings thusly: "All of the effects of alcohol are sort of amplified with age. Withdrawal is a little bit more complicated. Hangovers are a little bit more complicated."

Apparently, when you turn forty, your body becomes less interested in muscle and starts morphing it into fat instead—getting
you ready, one might assume, for that ice-floe push-off into the
dark, cold sea. But fat doesn't absorb or metabolize alcohol nearly
as well as muscle does. And at the same time, you'll be losing other
useful things, like body-water level and also dehydrogenase—that
enzyme that breaks down alcohol. So, if you somehow survive a
few more years and keep on drinking, theoretically you'll have
a lot more unabsorbed, unmetabolized, undiluted alcohol floating around in your body for a heck of a lot longer. And by the
time it's broken down, that stuff is going to hurt. It is worth mentioning that women, even if very fit, tend to have more body fat
and lower levels of dehydrogenase—a fairly valid reason why
(despite the suspicions of Dr. Stephens) they may in fact suffer
worse hangovers than men, even with the same BAC.

Then there's the liver. No matter its original gender, the organ
will enlarge as it ages, while both blood flow and the number
of hepatocytes (the cells that make it function) decrease. And
so—much like jockeys, governments, bearded dragons, concept
albums, ficus plants, skinny pants, fire ants, helium balloons
in a vacuum, Elvis Presley and this analogy—your liver grows
increasingly less efficient the bigger it becomes.

That previous sentence, it should be noted, was partly a
product of crowdsourcing. Unable to pinpoint the most appropriate simile, I asked some friends on social media for help. And
research suggests this might be something I'll need to start doing
more often, since apparently, our brains become less efficient over
time as well. Recent studies show that the connections between
neurons slow down as you age, turn downright sluggish if you
add a bit of booze and practically lobotomize you upon withdrawal. Did I mention I just turned forty?

ALTHOUGH BEGRUDGINGLY INTERESTED in my failing mind and liver, I am, of course, still on the hunt for a hangover cure. But it turns out Dr. Stephens doesn't believe in such a thing. "I don't think it's a suitable thing to study," he tells me, "in my very humble opinion."

Stephens is among those who see the hangover as a necessary phenomenon—vitally helpful to the health of society and ourselves. But of course, I still have to ask: "What can we do in the moment, to at least lessen our pain?"

The doctor shrugs. "Aspirin? Water? A good English breakfast? For anything more, you'd have to ask my kids." And so I thank him, wish him luck on his new book and hurry to catch a southbound train.

AND IT IS there, in the bar car of the train, that things start coming together. I am thinking about where I'm going, what I should order to drink and the workings of my aging body . . . when, just like that, my overworked brain starts doing it: *thinking divergently*—linking seemingly unrelated concepts.

The train speeds through a series of tunnels and I stare at the flashing window, making disparate, shape-shifting connections: a train station turning into a bar, a town revealed as a movie set, and then weirder ones: townsfolk turning into aliens, statues morphing into robots, public gardens blasted to smithereens . . .

Sure, a part of me wonders if this is just another method of delusion—a corruption of executive function causing thoughts to collide at rapid rates, a sorry state of intense unknowing, fostered by always being hungover. But the rest of me orders a nut-brown ale and plans my route ahead: it winds through the doors of twelve pubs, to the bottom of twelve pints—all the way to the World's End.

🍾🍾🍾

THE CORNETTO TRILOGY

The Cornetto Trilogy is a trio of films by three British best friends: Edgar Wright, Simon Pegg and Nick Frost. The first, *Shaun of the Dead* (2004), is a zombie film in which the characters played by Pegg and Frost awake on a Sunday morning into an undead apocalypse and take an awfully long time to realize it. Pegg, in fact, goes all the way to the corner store and returns with a red Cornetto ice cream (Wright's hangover cure from his college days) without noticing—because of his stultifying hangover—that his neighbors have turned into zombies. When they finally get what's going on, Pegg and Frost attempt to rescue their loved ones from various corners of town, then barricade themselves in their local pub for a bloody, ill-fated showdown. To this day, *Shaun of the Dead* is the only film that my parents, sisters and I have all watched together in an actual movie theater.

In *Hot Fuzz* (2007), which riffs on buddy cop flicks and so much more, Frost picks up a blue Cornetto for his police partner Pegg. During publicity for the film, an interviewer asked if they were doing a trilogy of Cornetto-based films. Wright answered that yes, they were: "Three Flavours," much like Krzysztof Kieślowski's deep and disturbing Three Colours trilogy. And then he felt obliged to make it so.

In *The World's End* (2013), Pegg, Frost and three high school buddies return to their hometown of Newton Haven to finish what they'd attempted twenty years ago—the Golden Mile: a one-night crawl of twelve pubs, with a pint of beer in each. Back then, they'd been thwarted from reaching the final post—the World's End—by various drunken mishaps. And this time seems even more doomed, since Pegg's character, Gary King, is fresh from a stint in rehab and only able to convince the others to go along with him by pretending his mother just died. Meanwhile, Frost's character, Andy, has long-since sworn off alcohol—mostly due to an accident caused by Gary, his former best friend.

"I haven't had a drink in fifteen years," he tells Gary.

GARY: *You must be thirsty.*
ANDY: *You have a very selective memory.*
GARY: *Thanks!*
ANDY: *You remember the Friday nights; I remember the Monday mornings.*
GARY: *Exactly! That's why we're going on a Friday night!*

But their biggest obstacle, as it turns out, is that the townsfolk of Newton Haven have been replaced by robots controlled by aliens trying to take over the world.

So that is part of what you need to know. But also this: *The World's End* was filmed in a seemingly unfitting, but in the end, bizarrely apt location, or rather a couple of them. Letchworth Garden City and Welwyn Garden City were the first "garden cities" created as part of England's idealistic New Towns movement around the turn of the twentieth century. Part of the newness involved a ban on selling alcohol in public houses—or pubs, for short—a ban that remained in effect until 1958.

The World's End, therefore, is the rendering of an apocalyptic pub crawl filmed in the one English town (or, rather, two English towns) where such a thing is barely possible—a belt of new-age British communities with more public gardens than public watering holes. With at least ten thousand other perfect locations scattered across England—each with its own actual Golden Mile of picturesque pubs to choose from—it was downright weird to pick the garden cities. And no doubt the location scouts had to widen their scope to encompass both, just to *find* a dozen pubs.

Even then, they apparently couldn't quite do it; among the ones appearing in the film, one is actually a restaurant, one a vacant shop, one a cinema and one a railway station.

It is for these reasons, never mind the inherent obstacles of downing twelve pints in twelve pubs, that the website Movie-Locations.com informs its readers plainly, "You won't be able to recreate the legendary Golden Mile crawl."

So that, of course, is what I plan to do.

TWELVE PINTS IN TWELVE PUBS (THE FIRST POST)

In the movie, it is the First Post, but in real life—here on the edge of Welwyn Garden City—it has always been the Pear Tree. And at midafternoon on a brisk but sunny day, it does look like the kind of place a festive partridge might roost—lush hedges and what I can only assume is a big old pear tree gently guarding gabled roofs and brick walls. From outside, at least, it is a classic, fine-looking pub.

I have come here in the various guises of beer drinker, film-goer, backpacker and researcher. My hope is that by attempting to recreate, pint for pint, a pub crawl from a movie in which middle-aged men attempt to recreate a pub crawl from their youth, I might gain some insight into how said movie's most dire warnings—about the idealization of youth, the glorification of booze, the homogenization of drinking culture, the brainwashing of consumers and the advent of Armageddon—pertain to the equally important conundrum of hangovers.

To tackle all this, I have enlisted a partner for the mission who is not only a knowledgeable local, trusted ally and Cornetto Trilogy fan, but is also half my age. As long as we can remain at least somewhat focused, we should be able to get to the bottom of something more than a dozen glasses. At the very least, we'll get to see who fares better—the forty-year-old veteran or the twenty-year-old hotshot. I have also acquired every known hangover remedy available at Boots pharmacy.

On entering the Pear Tree, I see that Tom has beaten me here—perched at a high-top in the corner, two full pints in front of him. His smile is as I remember: half-bashful and half-proud, with irony and slyness at the edges. It is good to see him—especially after traveling so long on my own.

I know Thomas Dart through his father, Jonathan, who is a British spy. Sure, Dart Sr. will insist he's just your average civil servant with lighting-quick reflexes, five pounds of skeletal titanium in his body, fluency in a dozen languages and perfect hair, but his name is Jonathan Dart, for crying out loud; that's even more Bond-ian than James Bond.

I met Jonathan Dart ten years ago now, when his official title was British consul general to Canada—no doubt the Queen's way of saying thank you for his previous postings in Korea and South Africa (where he was run over by a military jeep—twice). We became friends, went on a few adventures together, and then eventually he moved back to England, or—to be more precise—Letchworth Garden City.

Our Man Dart (now head of dealing with Libya or some such thing) is tied up in London until later, when he plans to meet us at the fifth or sixth post. But really, his son Tom is perfect for this mission. True, he may not yet have the same skill set and know-how, but he's inherited his father's spirit of adventure, devotion to detail and willingness to commit to a challenge, no matter how ridiculous. I sit down, and we knock our glasses together.

"Have you talked to the barkeep yet?" I say, looking around. There's an old-timer in a gray suit drinking alone at a table behind us, and four big guys with a somewhat surly feel to them at the bar.

"Well, I *did* order these pints . . ." he says cautiously.

"We'll talk to him after. So, what's the plan?"

In the opening voiceover of *The World's End*, Simon Pegg describes it as such: "A heroic quest: the aim to conquer the Golden Mile—twelve pubs along a legendary path of alcoholic indulgence: the First Post, the Old Familiar, the Famous Cock, the Crossed Hands, the Good Companion, the Trusty Servant, the Two Headed Dog, the Mermaid, the Beehive, the King's Head, the Hole in the Wall . . . all before reaching our destiny: the World's End!"

But given the true terrain—spanning twenty miles rather than one, with caveat pints scheduled in for posts that aren't actually pubs, a train trip partway through, as well as a precisely timed rendezvous with Her Majesty's liaison to Libya—our plan has required a bit more planning.

Tom has everything mapped out on his cell phone, which I haven't been able to reach all day for lack of reception—and also, the battery is almost dead. I show him what I've got in my knapsack: a shopping bag containing all the products a Boots pharmacist eventually admitted are intended to help with hang-overs (even though British law forbids them to be marketed as such), including Recovery Effervescent Powder (a sort of Alka-Seltzer with Tylenol), Re-Energise tablets (a version of Berocca's fizzy orange Aspirin tablets) and Digestion Relief Milk Thistle capsules; then there's my trusty laptop, some clothes, a recently forged fireplace poker and three leftover bottles of mead—surely everything we could possibly need for dangerous drinking and the advent of killer robots.

From my research so far, the milk thistle is the only one that may help in advance, so we each take a couple pills with our pints and knock our glasses together again. "Let's go talk to the man," I say.

"Okay. But maybe put the fire poker away first."

The barkeep's name is Martin. He's been pouring ale here for seven years, and he takes a brief minute to warm up—from bristly, to suspicious, to a wisecracking tour guide. It is a transformation we'll see again and again over the next many pints—a few of them on the house—until Tom will decide he's going to pretend to be a Canadian journalist in every English pub for the rest of his life.

In the film, Simon Pegg's character, Gary King, is under the misapprehension that his self-bequeathed high school nickname, "The King," has followed him into adulthood—and that upon their return, he'll get a homecoming worthy of the title. But at the First Post, no one—not even the bartender—remembers him. And then the joke at the second post—the Old Familiar—is that, once inside, it is identical to the first.

"They went out the main doors, then came in the side," says Martin, pointing to both entrances. "Movie magic, I guess."

I have a short list of somewhat boring questions I've decided to ask as we traverse the Golden Mile, and Martin shrugs his agreement to answer them—thus establishing the main themes to be repeated by bartenders throughout the night.

Q: *Have other people done what we're trying to do?*
A: *Oh yeah. Lots.*
Q: *Really?*
A: *Sure. Though I can't really think of who now. Students from the uni sometimes try. And I think there were a couple Aussie blokes. I don't know how far they got. But then, it can't actually be done, can it?*
Q: *I guess we'll see. What did you think of the movie?*
A: *Oh, it was rubbish. I mean, it started off good, with all the pubs and whatnot, but then it just went crazy balls.*
Q: *What do you recommend for a hangover—besides "don't drink" or "don't stop drinking"?*

A: *I'd say loads of water and a proper fry-up.*

Q: *Last question: What do you see as the future of British pubs?*

A: *Well, I don't know about robots and all that, but come back in a few months and this one here will be an Indian restaurant.*

Q: *Really?*

A: *That's right. It was here before the town, hundreds of years, and they sold it just last week.*

THE CLOSING AND OPENING OF THE MODERN ENGLISH PUB

The Industrial Revolution changed pubs as much as it changed everything else. All those small-town weavers, tanners, millers, butchers, bakers and candlestick makers pushed out of work by factories, then forced to take jobs on the assembly lines, gathered exhausted and defeated along the bar at the end of the day, and in larger numbers than ever.

Where once they'd drunk slowly, sweating it out in forges and fields, now they did it all at once, hard and fast, in an ever-expanding litany of public houses. And so, on the front lines of industrialization, a much more precarious awareness of hard drinking and the hangover arose, especially in regards to class.

As Clement Freud observed, "The working class victim gets very little sympathy on the 13 bus to the factory gates; the banker is shown understanding and envy by his colleagues, which is all the stranger because the manual worker will not work that much less well on the morning after, while the banker's creative output is likely to be nil."

With the advent of World War I, attention to worker productivity grew from the already weighty gaze of wealthy business owners to that of a whole empire. In an effort to curb such transgressions as drinking before the start of the morning shift in a munitions factory, leaving the line early for a pint or pounding them back late into the night and then showing up with a hangover, pubs could open only from noon to three thirty, and again from six thirty to eleven. And so it was for most of the next century.

"For generations, and by successive governments," wrote Andrew Anthony in the *Guardian*, "British people have not been trusted with a drink after the hour of 11 p.m. In the process, the nation's youth, to say nothing of its more mature citizens, have grown up under an effective curfew and a semi-Prohibition. In these rushed and draconian circumstances, it's no wonder that we have developed an unhealthy relationship with booze."

If you're to follow popular wisdom, the rise of modern binge drinking (and neo-drunken hooliganism) sort of sloshes into the sinking of the British Empire and is often blamed on the dual uncivilized forces of a condescending closing time and immature, sometimes violent, drinking habits. The result, as former Home Secretary Jack Straw put it, was people "hitting the streets—and sometimes each other—at the same time."

By the end of the millennium, you'd have been hard-pressed to find a Brit—whether in public office, the media or even just sitting in a pub—who was unready to ridicule and lament the unsophisticated intoxication of their fellow citizens. "Other countries tend to be ashamed or mortified by drunkenness," observed epidemiologist Annie Britton in her study of drinking among British public servants. "The UK is unique in glorifying it with programmes such as the appalling Booze Britain. And our drinking habits have got worse."

When, in 2003, Parliament finally announced the dismantling of Britain's harsh pub laws, it hoped to promote a civilizing influence on the nation's boozing habits; others foresaw an alcoholic apocalypse. And the most attentive scribes of England rose to the occasion with gleeful, yet remorseful eloquence.

"We drink. We pee. We drink. We dance. We drink. We eat. We vomit. No one else does this for fun," wrote Giles Whittell in the *Times*—just months before the advent of twenty-four-hour British bar licenses.

In those same few months before the English pub would never again be closed, Andrew Anthony also examined the finer points of British intoxication:

> *Sometimes, especially on Fridays and Saturdays, it can seem that sobriety is our national enemy, an implacable foe that needs to be fought on all fronts at all times, or at least until the pubs close. . . . Whoever you want to accuse—including a hysterical media that both celebrates and demonizes drunkenness—the inescapable fact remains that the British, by and large, like to get drunk more than they like drinking. That is to say, we take a utilitarian approach to alcohol: what's the point of consuming it unless we end up rat-arsed?*

Or, as Gary King and the lads debate as they embark on the Golden Mile:

> *—What is this? Why are we even here?*
> *—We're here to get annihilated!*

TWELVE PINTS IN TWELVE PUBS (THE NEXT FEW POSTS)

In the movie, the joke of the Old Familiar is twofold: not only is the interior identical to the previous pub, but no one recognizes them here either. In real life—at least from the outside—it is the

Doctor's Tonic, a pub for which its fictional name now rings even more ironic.

You can picture how it once might have been: a big old public house containing all the dichotomies of British boozing: at once comfortable and dubious; a cozy sort of mystery with hidden rooms, dark corners and limericks carved into the table legs; a place where strangers were greeted with quiet stares or pints on the house, and sometimes both; where the plates were chipped, but a hundred years old; where the glasses didn't match, but were rimmed with gold; where the barmaid was a bitch, but the most honest person you'd ever known. An archive of all the secrets, dreams, lies, piss-ups, punch-ups and homecomings of a thousand different families. And now it is none of that.

Today, the good old Old Familiar/Doctor's Tonic is a refurbished, very clean establishment with a large patio, full parking lot, prompt young servers in matching collared T-shirts, a half dozen flashing arcade games and a Top Forty playlist. In short, it is a well-run, successful business, and a lot of what is wrong with modern British pubs—at least according to Kingsley Amis.

In his introduction to *Everyday Drinking*, he explained it like this:

> *Fifteen or twenty years ago, the brewing companies began to wake up to the fact that their pubs badly needed a face-lift, and started spending millions of pounds to bring them up to date. . . . The interior of today's pub has got to look like a television commercial, with all the glossy horror that implies. Repulsive "themes" are introduced: the British-battles pub, ocean-liner pub, Gay Nineties pub. The draught beer is no longer true draught, but keg, that hybrid substance that comes out of what is in effect a giant metal bottle, engineered so as to be the*

same everywhere, no matter how lazy or incompetent
the licensee.

Another twenty years have passed since Kingsley's lyrical grouchiness, and now, thanks to our hipster-craft-beer-and-foodie culture, the food and beer aren't actually so bad. At the same time, however, places like these have managed the near-complete homogenization and franchising of the little things we loved, even if we didn't know we loved them. As the boys coming back to Newton Haven explain, "It's all part of the nationwice initiative to rob small, charming pubs of any discernable character."

"Starbucking, man. It's happening everywhere."

Alex, the manager of the Doctor's Tonic, brings out the Old Familiar sandwich board for us to take a photo, and suggests "a Bloody Mary and a greasy fry-up" for the morning after. I neglect to ask him about the future of the British pub. And then, on arriving at our third post, I'm afraid we may have found it.

In the film, it is the Famous Cock, and when they were shooting, it was actually the Cork. But by the time Tom and I show up, it has become the Two Willows—part of the Stonegate Pub Company's line of "Classic Inns," each "uniquely" branded to some bucolic-sounding aspect of the terrain, even if it had to be carted in.

So, whereas the Pear Tree, soon to be an Indian restaurant, sits beneath an actual ancient pear tree, the Two Willows' front door is framed by two planters, containing two cocky little saplings that may, one day, become two actual willow trees. The horse-after-the-cart aspect to all this becomes more insidious inside, where, surrounded by dark, shining mahogany, glistening leather, sepia prints in glass frames and wall murals of Victorian-era Garden City street maps, the metaphor morphs to that of a transplanted garden snake consuming its own tail. Because now it is becoming clear: really, this is a "theme" bar. And the theme is "Local British Pub."

As the remarkably nondescript man behind the bar is pouring our pint, I ask if he happens to be the manager.

"I am one of the team leaders," he says with a nod, leveling the foam.

We introduce ourselves and tell him what we're up to, and he points us toward a window separating two booths in the back, the glass tastefully engraved with THE FAMOUS COCK. This is where Gary King is finally recognized, as the barkeep in the film points to a taped-up photo with the caption "Barred for Life." But still, Gary manages to make his quota, quickly gulping down a makeshift pint from the dregs left out on the patio.

The team leader, whose name is Bobby, places our perfectly poured pints on the bar and earnestly answers my questions. The film was "good for the local economy." The English pub is on its way out, and that is "sort of a tragedy."

And the best treatment for a hangover?

"That," he says, looking us in the eye, "is an orange-flavored iced lolly."

They are such perfect, perfectly creepy, answers that now I am sure: if it turns out that no one else in town is actually an alien robot, at least there is Team Leader Bobby.

THE NEXT POST, the Crossed Hands, is the turning point in the movie. The boys learn that Gary's been lying about his mum; Gary fights an alien robot in the guise of a young hipster in the bathroom; they *all* fight alien robots in the bathroom, decapitating them with their severed robot limbs, blue blood spraying everywhere. Then they realize that *all* the citizens of Newton Haven may now in fact be robots, and Andy falls off the wagon—downing five quick shots to catch up.

In real life, it is the Parkway, and the first bar we've been to so far with no hint of refurbishment, let alone any Starbucking. The outside has a working-class sports-bar feel to it, and the inside is one big, open room with bare rafters and wood paneling, kind of cabin-like. We're drawn to a large booth on a riser at the back— and later, looking at the movie again, I'll realize this is precisely the table where the lads hunkered down.

Before us are two pool tables, both with people playing on them, while the other half dozen patrons are sitting at the bar. There's something both insular and comfortable about the place, like these are actual regulars who kind of like it here. We leave our bags and go to the bar.

Behind the bar are a young guy and gal. He is good-looking and gangly and is hammering a shelf into place. She is short and cute and seemingly not very comfortable with the kegs, or my first question. "They want to know about the beer " she calls out to her colleague.

"What about the beer?" he says.

"It's okay," I tell her. "We'll take one of each."

"It's okay!" she calls out.

We take the pints back to our table and, when he's done hammering, the tall guy comes over and we chat. He's charming, if a bit hard to get a read on. They both are. Their names are Conner and Jade.

"Hey!" calls Conner to Jade, picking up on the least important of my questions. "What are our job titles?"

"I don't know," answers Jade seriously. "We work at the bar."

"There you go!" he says, tapping the table with a grin. "Oh, and I've got a good hangover cure. Remind me to tell you later." Then he heads back over to the bar.

"I like this place," I say, and Tom lets out a laugh. "What?"

"Look around," he says. "It's basically a small-town Canadian bar."

And cor blimey, he's right. Who'd have thought I'd take the

makings of a Canuck roadhouse over a half-dozen English pubs, especially on this side of the ocean? But this, so far, is the only one that feels real.

Conner comes back with shots on the house. They're like little lava lamps—red and green splotches suspended in swirling yellow. "Traffic lights," he says.

We say thanks and ask what's in them. He's about to answer, then lifts a finger, goes back to the bar and returns with something scribbled on a piece of paper. "Some rum, some Archers, some Warninks and then some red stuff."

"Red stuff?" says Tom.

Conner holds up his finger again and goes back to check. "Yep," he calls out, "red stuff!"

We hold up our traffic lights and shoot them back.

Conner and Jade think they might have seen *The World's End*, but can't be rightly sure. They seem to remember a couple of guys doing what we're doing—maybe from Australia. They think the future of the British pub doesn't matter much. And finally, Conner has his hangover cure for me to write down and put in my book.

"Here's what you do," he says, like it's a heavily guarded secret. "Before you go to bed, drink a pint of water. Then, in the morning, you take two paracetamols and then . . . *then* . . . you eat a big classic fry-up!"

"So," I say, diligently writing it down, "water, Aspirin and breakfast . . ."

"That's right," says Conner, perfectly deadpan. Then he gets back to hammering.

"Our Conner," says Tom. "He's a right geezer."

But me being Canadian, and our Conner being much younger than me, I have no idea what the hell Tom is talking about.

"Not like how you guys use *geezer*. It's more like . . . well, like a guy that everybody knows, and kind of shakes their head—fun, but a bit dodgy. It's hard to explain."

"Like Gary?" I say.

"Who?"

"Gary King. The hero of the movie . . . the one we're kind of *in* right now."

"Oh, yeah," says Tom. "He's a *right* geezer."

In fact, if I'm getting it right, Gary King is kind of the ultimate geezer—a man so adamantly dodgy, weirdly charming and averse to authority that he has fought against the progress of time itself—stuck for all those years in a moment when his rebellion seemed to make sense and they took on the Golden Mile.

The World's End, after all, isn't only about a pub crawl and robots. The idea of alien superpowers Starbucking the universe—trying to refurbish Earth into a uniform, brand-specific, safe and homogenous planet, to save us from ourselves—smacks of an imperious imperial government that treats its citizenry like irresponsible teenagers and shuts the bars an hour before midnight for their own damn good. In such a world, the most imperfect citizens become its only hope. And the only proof Gary King is able to provide that *he's* not a robot are the fresh scars on his wrist—the ones that put him into mandated rehab. There is a lot going on below the surface.

THE NEXT POST, the Good Companion, doesn't exist in real life as more than an abandoned storefront. But we decide the traffic lights, both liquid and not, have taken care of that. So only a half dozen pints to go as we miss our train, haggle with a cabbie, then head toward Letchworth—the other garden city—and its six remaining posts.

In the movie, passage to the next few pubs is played out in a kind of montage set to Jim Morrison's cover of Bertolt Brecht's "Alabama Song" (*Show me the way / To the next whisky bar / Oh, don't ask why / Oh, don't ask why*). The sometimes-robotic nature of mindless alcohol consumption is played up here, much like the zombification of hangovers in *Shaun of the Dead*, as the boys decide that the only way to avoid suspicion is to stick to the original plan, one pub and pint at a time.

In real life, when you drink five pints and a shot, then take an English cab through twilight turning to night, from one garden city to the next, whether you're twenty or forty, at the start of your drinking life or halfway through, something starts to happen; you notice, at the very least, that now you're a little bit buzzed—perhaps even getting drunk.

Since we're supposed to be on a train, our rendezvous with General Dart is at the train station—which in the movie is the Beehive. There's no actual pub here, but there is a nice little Italian restaurant that serves pints. And the owner—a dashing, Old World kind of guy, has one with us.

"Silly movie," he says. "But they were really nice blokes."

I'm about to ask him about hangover cures and the future of Italian restaurants in British train stations, or something like that . . . but then, suddenly, the always-stealthy Jonathan Dart appears, clapping us on the back.

"Three drunk men!" shouts Tom.

It is a reference to a scene near the end of the movie, where Gary and Andy find themselves underground, arguing with the invisible head of the galactic corporation on the grounds that earthlings *like* being the way they are.

"And therein lies the necessity for this intervention," says the disembodied voice of Bill Nighy. "Must the galaxy be subjected to an entire planet of people like you? . . . You act out the same cycles of self-destruction again and again."

"Hey! It is our basic human right to be fuck-ups! This entire civilization was founded on fuck-ups! And you know what? That makes me proud! . . . What is it they say? 'To err is . . .'"

"'To err is human.'"

"'To err is human!' So . . . errr . . ."

"You do not speak for all humanity. You are but two men. Two drunk men . . ."

"*Three* drunk men!" calls out their last human friend on the whole damn planet, rappelling down from the surface of the Earth.

"THREE DRUNK MEN!" I echo back, Tom and I raising our glasses.

But Jonathan Dart—direct from managing the Western world's interests in the Middle East—appears incredibly sober, so clear-eyed, even-keeled and properly enunciating in his pristine jacket and tie, that just like that, by way of relativity my blood alcohol level spikes, and Tom and I appear suddenly off-balance.

"Quick!" I say. "You have to have a pint!" But then, the manner in which Jonathan Dart orders a pint of cider is so ridiculously cool that I ask him to do it again, so I can record it on my phone. And now Tom and I better have another half pint too, because really, when it comes down to it, those traffic lights can't count for a whole post . . .

After a couple more hours, and three more posts, the Earth has become tilted and we three drunk men are weaving our way down the road, singing, belching, laughing and arguing about nothing. The World's End is much farther than I'd ever imagined—all the way to the edge of town. And by the time we get there, it is closed.

In the film, the World's End is closed by the time they get there too, but the boys crash through the front wall in a truck while fighting off robots as well as each other. Finally, Gary King breaks free and makes it to the draft tap for a final pint. But when he pulls it, the ground opens up and they descend to an alien lair.

In real life, we shrug, turn around and head for the Dart family home. "Here's the plan," I say, as we weave back down the road. "We get to the house, get three tall cans from the fridge, go find a place we won't be waking anybody up, and bingo: twelve full pints!"

"That's an excellent plan!" says Tom. "What do you think, Dad?"

"I think," says his dad, "that I can't reach twelve tonight either way, and have to work in the morning, so I'll be going to bed. Oh, and Tom, I happen to think you've had more than your fill, but I also think you're an adult now, so truly it's up to you."

"Yes!" says Tom. "It is an excellent plan!"

"But also," says Jonathan Dart, stuffing his tie into his pocket, "even in the film, didn't they only get to eleven—then they dropped down to that underground lair?"

BENEATH THE EARTH, Gary and Andy find their footing, now facing some glowing alien tribunal. Spiraling balconies rise above them, holding the robots and shadows of Newton Haven townfolk.

"We can offer attractive incentives for those who willingly comply," explains the voice of Bill Nighy. "The chance to be young again and yet retain selected memories. Isn't that something you'd like? Something you've always wanted?"

At that moment, a light bursts on and there—standing before them—is forty-year-old Gary's twenty-year-old self—the embodiment of his long-ago hopes, dreams and all his roguish optimism. "Oh my God," Gary gasps, "I'm so cute!"

AND NOW, HERE are we, on one of the public lawns of
Letchworth Garden City, where drunken twenty-year-old Tom
grew up. It is a big lawn—a shrub-lined fence on three sides and
a tree in the middle, against which my forty-year-old back is leaning while I look at the stars and drink my dozenth beer.

I am listening to Tom, who is overwhelmed and full of it
all. It is shining out of his brimming eyes as he laughs and
cries. He is in love and verging on heartbreak, scared and
brave and ready to go—to fight through jungles, cross deserts
and leap into the waves off the sharp edge of the world. What
I wouldn't give for that: my buzzing, brave heroic self of long
ago. I remember it so clearly, yet impossibly—like a dream
of flying.

CUTE YOUNG GARY King walks toward his older self. "Allow
me to carry your legend forward," he says. "Let the man you've
become be the boy who you were."

Strung-out, washed-up, cut-up Gary looks at his golden self.
And for a moment, they are fixed, their eyes softening, smiling . . .

"Nah," he says, then cuts off his younger self's head with a
samurai sword.

He shouts to the tribunal, "There can be only one Gary King!"
and punts the severed head into the surrounding darkness.

TOM IS OFF and into the shrubs, and I wouldn't want to be him
now. After a while on his knees, he gets up and starts stumbling
around the garden square, his right hand bouncing off the fence,
spinning against the spinning.

And me, I sit in the middle, against the tree, watching him go
round and round, talking him down and then back up. I finish

my beer, then start on Tom's twelfth—my thirteenth—as he gets rid of most of his.

I have a sense of protectiveness, empathy, envy and nostalgia all at once. And while I drink, I look at the stars, a million golden miles away.

A WHOLE NEW MEANING

When the Starbucking network of meddling aliens finally (spoiler alert!) gives up on Gary King and the rest of humanity, their departure sets off a blast that destroys all technology and causes the apocalypse.

"That morning," says Andy, speaking from Earth's future, "gave a whole new meaning to the word *hangover*. We decided to walk it off on the way back to London. But it didn't end there; it just went on and on and on. We all had to go organic in a big way. But if I'm honest, I'm hard-pressed to recall any processed foods I actually miss." We see Andy behind a fence, dressed in medieval road-warrior garb, a shovel in his hands, a goat at his side. And now something blows onto the chain-link: a green Cornetto wrapper. As if by primal urge, he leaps for it. But it is gone with the wind.

Gary King, who was in his element as a rebel teenager, has found his place again in the post-apocalypse—leading a tribe of thirsty, abandoned robots through the wasteland of humanity. In the final minute of the movie, a beef-faced bartender refuses to serve the refugee robots and demands to know their leader's name.

"They call me 'The King,'" says Gary.

He unsheaths his sword and leaps over the bar.

Fade to black.

🍾🍾🍾

YEARS BEFORE THE Brexit vote, England abolished a hundred years of legislation in the hope of becoming something they never were: European drinkers. It didn't trigger Armageddon, exactly, but neither did it change British boozing habits. In fact, it seems the opposite has happened. The supposed sophisticates on the continent have instead become barbarians.

"What has gone wrong?" wrote Jon Henley in the *Guardian* in 2008. "What has prompted France's youth to turn from sensible tipplers to full-on booze abusers?"

Étienne Apaire, who heads up an interministerial body aimed at combating drug and alcohol addiction, believes the phenomenon is part of a "globalisation of behaviour" in which teenagers everywhere are now seeking "instant intoxication" as an end in itself.

Or, as Andy put it to the big, faceless alien: "We humans are *way* more stupid and stubborn than you can even imagine."

TWELVE PINTS IN TWELVE PUBS (THE WORLD'S OTHER END)

By the time we get up, Jonathan Dart is already in London, trying to save the world from itself. So Tom and I, feeling a bit rough, take all that Boots had to offer and head back out to finish our mission. We walk bleary-eyed down the avenues, across the lawns and public gardens, through the neighborhood of Tin Town— where bombed-out Londoners were housed in sheet-metal huts during World War II, then decided to stay for generations after— across the railway on a caged-in overpass and all the way to the World's End/Gardener's Arms, where breakfast is served every day until noon.

We order Bloody Marys and some water and debrief the Golden Mile. I assure young Tom I won't write about him throwing up. But I can't in good conscience say he drank twelve pints

in twelve pubs, since I drank his beer while sitting against the tree.
So Tom orders a pint in addition to his Bloody Mary, and we start
to go over the posts, each by each—but each time we do, one is
missing. We get out our phones, ask for pens and paper, look at
the data as best we can, and then Tom slams down his beer.

"*The Two Headed Dog!*" he says. "We never made up for the
Two Headed Dog! We got too drunk and forgot about it!"

"So . . ."

"So we only drank eleven pints!"

"Well, actually," I say, "I had your last one. So really . . ."

"Oh shit! I only drank ten! Ten measly pints!"

So there we are: twenty-year-old Tom drinking two pints to
go with his Bloody Mary. And forty-year-old me drinking one
more—just to be bloody sure.

IN LOOKING AT the outcome of our mission, there are too many
mitigating factors to simply compare our hangovers. It does seem
fair to say, however, that age and treachery outdid youth and vitality when it came to holding our drink—or at least holding it all
down. And then, of course, his reverse peristalsis could have mitigated some of Tom's hangover symptoms. Either way, neither of
us appears obviously worse off than the other this morning. And
I'm starting to think that the milk thistle might have done us good.
For now, I'll put it in the column of things I plan to test some more.

In regards to my coinciding investigation on the spectrum of
popular hangover lore as offered by local publicans, that one seems
fairly conclusive. Except for Team Leader Bobby and his rather
surprising orange ice lolly, the overwhelming prevalence of "breakfast, water and an Aspirin" is as undeniable as it is uninspiring.

The lack of diversity and ingenuity is enough to make one long
for the days of Pliny the Elder—or better yet, Frank M. Paulsen.
A mere fifty years ago, after all, his anthropological pub crawl

through bars and roadhouses of North America resulted in a list of 261 different barstool suggestions, including parsley, parsnips, persimmons, prune juice, pickled herring, papaya, pepper on ice cream and performing cunnilingus in a very particular way—and that's just a small part of the letter *P*.

Meanwhile, Sarah Marshall, the journalist whose article first put me onto Paulsen's work, had much the same mundane results as I upon conducting her own modern-day barroom study on the other side of the ocean. "Was I asking the wrong people," she reflects, or has drinking culture "changed to such a degree that we had lost touch with the 'gigantic vortex of folklore' that had so obsessed Paulsen?"

And now, I guess, is the time to divulge what that self-proclaimed British "bad witch" had to offer by way of email:

> Hi Shaughnessy:
>
> Drinking plenty of water and maybe a couple of Aspirin is the usual recommendation as far as I know. Mind you, I have heard a few people say that a full English breakfast is a good hangover cure, and you could try that when you are here!
> the best,
> Lucya

Could it be that all of us, even proper geezers and bad witches, have already been thoroughly brainwashed—Starbucked into a realm of plastic menus and easy answers, with no fight left against the monolith of the morning after?

"Hey," I say, to our noonday bartender at the World's End/ Gardener's Arms. "What's *your* recommendation for a hangover?"

"I don't have one," he says, putting down our bill. "Never had a drink in my life."

THE WITHNAIL AWARDS: A PRESS RELEASE

From *The Lost Weekend* to *Dude, Where's My Car?*, the morning after has been a part of the movies since long before *The Hangover* made it so very epic. How many transitional scenes in all genres have begun with our hero waking, head buried beneath a pillow, one arm flailing for that out-of-reach alarm clock? It is as familiar a trope as the meet-cute or the slow-clap, and without it, some of the most well-timed fade-to-blacks would have had nowhere to go.

With that in mind, the Academy of Motion Sickness is pleased to announce the nominees for the International Withnail Awards, lovingly known as the Queasies. Broadcast whenever we happen to get around it to it, these awards recognize the vast array of achievement in the history of hungover cinema—and are, of course, named for the drunken, out-of-work actor played to perfection by Richard E. Grant in the 1987 classic *Withnail & I.*

This year, in recognition of the epic drinking game Quaffing with Withnail, whereby fans match the eponymous character drink for drink throughout the film (a total of nine and a half glasses of red wine, half a pint of cider, one shot of lighter fluid, two and a half shots of gin, six glasses of sherry, thirteen glasses of whisky and half a pint of ale) we are encouraging viewers to come up with their own drinking game for the broadcast. For example: a shot every time the host fake-hiccups, a presenter wears his tie untied or Nick Nolte accidentally stumbles onto the stage.

And now, we are pleased to announce the nominees:

WEIRDEST WAKE-UP. *Dumbo* (in a tree), *Sixteen Candles* (also in a tree), *Old Boy* (in an underground cell), *Fear and Loathing in Las Vegas* (wearing a lizard tail and a duct-taped microphone, surrounded by empty coconuts and condiment-smeared altars), *The Changeup* (in someone else's body), *Hot Tub Time Machine* (in the 1980s).

MOST DESTRUCTIVE MORNING-AFTER PLOTLINE. *Hancock* (hungover superhero causes $9 million in damage to downtown core); *Die Hard: With a Vengeance* (psychotic terrorist bombs city, ruins perfectly good hangover); *Shaun of the Dead* (zombie apocalypse coincides with Sunday morning); *End of Days* (standard apocalypse coincides with hungover Schwarzenegger.)

LEAST LIKELY ALCOHOLIC. Arnold Schwarzenegger (*End of Days*), Sandra Bullock (*21 Days*).

MOST CONSISTENT HANGOVER PERFORMANCE. Peter O'Toole (*My Favorite Year*), Matt Dillon (*Factotum*), Nicholas Cage (*Leaving Las Vegas*), Johnny Depp (*Pirates of the Caribbean: Curse of the Black Pearl*).

MOST DRAMATIC USE OF BODILY WASTE. *Get Him to the Greek* (vomit), *Trainspotting* (diarrhea), *Bridesmaids* (vomit and diarrhea).

BEST HUNGOVER DIALOGUE. *Anchorman* ("I woke up this morning in some Japanese family's rec room and they would not stop screaming"); *My Favorite Year* ("*Ladies* are unwell. Gentlemen vomit"); *Die Hard: With a Vengeance* ("Beer is usually taken internally, John"); *True Grit* ("I know you can drink whisky and snore and spit and wallow in filth and bemoan your life and station. The rest has been braggadocio").

Finally, in addition to this year's categories and nominees, we are pleased to announce the recipient of the Lifetime Queasy Award. Joining the ranks of Jimmy Stewart, Paul Newman and Jessica Lange, the great Jeff Bridges will be honored for his stumbling, mumbling morning-after turns in such classic films as *The Fabulous Baker Boys*, *The Fisher King*, *The Big Lebowski*, *True Grit* and *Crazy Heart*—though not, ironically, *The Morning After*, in which his character remained bafflingly sober throughout.

We hope that you'll be joining us for this year's Withnail Awards, and calling in sick the morning after.

PART SIX

THE HUNGOVER GAMES

IN WHICH OUR MAN IN SCOTLAND DRIVES A JEEP
BLINDFOLDED, SHOOTS ARROWS IN GALE-FORCE WINDS
AND ATTEMPTS TO "TASTE" RATHER THAN DRINK.
APPEARANCES BY MALT MASTER KINSMAN, SIR ARTHUR
CONAN DOYLE AND THE LEGENDARY MAX McGEE.

*There are natural born drinkers just as there are natural born
athletes who need very little honing, but as the cognoscenti of
either endeavor are well aware, a W.C. Fields or a Willie Mays
comes once to a generation, if at all.*
— GEORGE BISHOP

C HICANE!" SHOUTS MY GUIDE FROM THE PASSENGER seat. Never mind that his brogue is so thick I can barely tell when he's saying "left" or "right"—or that I'm slightly hungover, and very blindfolded—it is this word that makes me stop. Standing on the brakes, skidding through the mud, I turn my head and call into the darkness: "Chicane? Really?"

To my limited knowledge, the word is either a precise Celtic directive or a general racing term meaning alternating curves—either way, no help to a sightless, shaky Canadian who is driving a jeep on the wrong side of the Scottish Highlands.

"You aren't moving," he says. Or at least I *think* that's what he says. To test my assumption, I stand on the gas. There's no way my team is going to lose whatever it is we're attempting to do: Team Hangover versus Team Speyside, Realists versus Publicists.

When it comes down to it, we're all just drunker journalists, here for Glenfiddich's 125th birthday—a lush and classy three-day party at the Speyside distilleries, the rough-hewn yet majestic Highland home of Grant's, Glenfiddich and Balvenie. We've been pretending to nose and taste, when what we really know how to do is drink.

And now, this morning, we've been placed in the hands of some corporate team-building company, Scottish Highland–style—which means attempting to steer a jeep without sight, then launch arrows without skill. We were also supposed to try laser-guided clay pigeon shooting (whatever that means), but the guy in charge of it crashed his car on the way here and wound up in the hospital. Though necessarily concerned for his well-being, the effort to not make blindfolded-driving jokes at his expense has been exhausting for everyone.

But I don't think the German food writer who got in the jeep before me was even kidding when he said, "If you are blind, the other senses get to be heightened, ya?" He stood on the gas, turned the wheel, eased off and then inched through the Highland mud at about three miles an hour. As it turns out, if the windows are closed and you've only been blind for a few seconds, your other senses don't tell you much at all.

But I think I'm moving fairly fast—at least based on the sounds from Shrek over in the passenger seat. And I'm confident I can handle this, sight or no sight. Hell, I've driven a racecar on a ten-turn track during that pivotal moment between high and hungover—and right now, as it happens, I'm feeling relatively fine. Relative, of course, to how much I drank last night—which was really quite a lot.

When you're being offered some of the oldest, finest Scotch in the world straight from the barrel, you damn well take a hearty sip. And if you nose and taste and spit it out, well, then, I hope to Dionysius the devil will get you. The point is that this morning, when I should at least be carsick—spinning in the muddy dark with a humorless Scot spitting arcane instructions at me—I'm actually feeling pretty good.

"Out you go!" grunts my guide; at least, that's what it *sounds* like. I open the door, then take off my blindfold and see that we're still moving. So I pull the emergency brake while stepping out of the car. He's cursing as I close the door. I take a breath of the fresh Highland air.

♠ ♠ ♠

IN *TRAINSPOTTING*—Irvine Welsh's poetic study of Scottish debauchery—the boys, led by the hopeful Tommy, try to shake off the Glasgow bug by heading into the hills. The goal is a noble Celtic one, but the weight of their hangovers, combined with a crushing sense of societal discord, doesn't allow them to quite reach it. Danny Boyle's movie adaptation has them staggering toward the Highlands:

> —*Tommy, this is not natural, man.*
> —*It's the great outdoors. It's fresh air. . . . Doesn't it make you proud to be Scottish?*
> —*It's shite being Scottish. We're the lowest of the low. . . . We can't even find a decent culture to be colonized by. We're ruled by effete assholes. It's a shite state of affairs to be in, Tommy, and all the fresh air in the world won't make any fucking difference.*

Despite this immutable Scottish logic, the go-to medical advice for any manner of baffling ailment—exhaustion, depression, anxiety, lethargy, paranoia, delusion, nerves—was, for centuries, a bit of clear mountain air. And so it was for the coinciding of all these symptoms—such as found in the aftermath of a truly epic binge.

There are many scientific reasons as to why high heights may actually worsen the morning after. And those who have suffered severe mountain-climbing sickness tend to equate it, at least in physical terms, with the most crippling kind of hangover. Then again, low-lying hangovers—as one might find in Vegas or Amsterdam—can send you running for the hills an age-old attempt to find salvation. But is there anything to it?

In *The Wrath of Grapes*, Andy Toper relates an early-morning reconnaissance mission described to him by Officer George Farrow: "The more I climbed . . . the more my hangover disappeared. By the time I reached the Cameron Highlands my head was completely clear and my stomach no longer felt queasy. I was ready for lunch and even a few more drinks—the thought of which would have made me throw up earlier."

As Toper assesses: "The simple explanation of this phenomenon might be that climbing a mountain can cure a hangover (although it is generally accepted that altitude increases the effect of alcohol). But then nothing about this affliction is ever that simple."

IT IS NOW almost noon, with near-hurricane winds—so, clearly, a perfect time for amateur archery.

"Who here has shot an arrow before?" says the younger, less broguey, of our two remaining guides. This one, it seems, has a surplus of energy, and spiky hair so full of gel as to be impervious to whirlwinds. "None of you?" he shouts, with incredulous glee.

No one budges. Our eyes are streaming and we're trying to keep our flapping clothes from tearing right off. But Spikey is opening his mouth again, so I shoot up a hand and take one for the team. "I have," I say.

"Ho there, Braveheart! What shall we call ye?"

Instead of pointing out that he just called me Braveheart, I give up my name.

"Okay, then, Shocknessy!" He hands me a bow. "Let's see what ye can do!"

Historically, I cannae do much. But it seems this profession of mine tends to land me in places where I'm meant to learn strange things. Just last month, for example, for a fatherhood column I write, I found myself at a workshop on "Archery and the Undead."

"What have you happened to shoot before?" asks Spikey, presenting a quiver of arrows. That day, we'd fired on scarecrows dressed as zombies, with pumpkins for heads.

"Pumpkins," I tell him.

Spikey hoots. "Give Shocknessy a hand!" he says, as if I came here to brag about it. "Our man killed an inanimate vegetable!"

Holding the bow, I nod a gritted-teeth smile at the stupid redundancy. Normally, I'd be fine with the piss being taken out of me, but this wind is blowing like hell, I'm the one doing Spiky a favor, and I'm actually quite proud of killing that pumpkin-headed zombie/scarecrow. Much more hungover that day, my arrows kept tumbling sideways into the mud. Then, nearly at the end of my quiver, I'd realized something: I'd been nocking my arrow on the wrong side of the bow the whole time. I drew one more—and stuck it right between those undead pumpkin eyes.

So Spikey doesn't know who he's dealing with: a man who's done far more ludicrous archery, in a much worse state of hangover; a soul well practiced at pretending to know what he's doing. I lick my finger to check the slant of hurricane, narrow my eyes on the balloon lashed to a far-off bale of hay, angle the arrow up toward a long, impossible arc. And then I let it fly.

AGAINST ALL ODDS

Every morning across the globe, millions of people fail miserably at whatever they were supposed to do, or don't show up at all.

But then there are those times when the guts, tenacity and desperation it takes to overcome a hangover can spark surprise success—and maybe even greatness. Sure, writers, rock 'n' rollers, actors, steelworkers and so on get right up against it all the time, maybe every day. But the stakes and stages are rarely as big as they are for big-time athletes. Combine intense physical competition and team and public expectation with the intrinsic idea of pushing oneself to the limit, and the most hallowed ground of heroic hangovers is historically found on the fields and pitches of professional sports.

British Formula 1 champion James Hunt had a lust for life and indifference to the sport at which he excelled. Race cars were simply the fastest way to get to all the best parties. In fact, it was drinking with Hunt that Stan Bowles blamed on an epically disastrous performance at his own Hungover Games.

The seemingly mythical, and sadly just deceased, Bowles was the Hunt of British football, with a bit of Shane MacGowan thrown in. His lack of foresight was as legendary as his drinking and gambling. This was a man who signed sponsorship deals with two different shoe companies and dealt with the fallout by wearing one on each foot.

Then, in 1974, he won the dubious honor of achieving the lowest score ever on *The Superstars*—a televised competition that pitted elite athletes from different sports against each other in a series of Olympic-style events. To quote a summary of his underachievement in the *Guardian*: "He couldn't complete a single length in the swimming, failed to clean and jerk the weights, lost 6–0, 6–0 to [Welsh rugby star] J.P.R. Williams in the tennis, was engulfed by a wave in the canoeing, and shot a table in half in the shooting."

Explained Bowles about missing the target so drastically: "I'd been out with James Hunt the night before and I had a right hangover and it just went off. A hair trigger, you see?" But also, that seems to be how most of his days happened to go. As Bowles wrote in his autobiography, "Everywhere I went, it ended up in chaos." Except when he was on the field. Then he moved like a

Jedi, the ball bending with the force of his skill. And if too far
gone, he could always collapse on the bench.

That's the thing about team players rather than individual
antiheroes left to their own devices: hangovers can often be
sidelined, absorbed by coaches, teammates and a solid piece of
wood. But it's worth jumping over the ocean, to that other kind
of football, for an instance of the opposite: the first-ever Super
Bowl, the most epic telling of which is by Shaan Joshi, an East
Indian from Indiana who wrote in the *Prague Revue*: "Wherever
men meet in skill and competition legends arise. . . . Sometimes
those same legends show up at the team buffet the morning
of the championship game reeking of booze and no sleep. . . .
Sometimes that legend comes off the pine and dominates with a
raging hangover. Sometimes legends have a name. This legend's
name is Max McGee."

The first-ever Super Bowl was played in Los Angeles,
between the Kansas City Chiefs and the Green Bay Packers,
whom McGee played for. But he'd been on the bench most of
the season, and there was no reason to think this would be any
different. Plus, L.A. was his town, so he was going to hit it—
curfew be damned.

As Joshi puts it: "Max was living high. Max was living
right. What the hell did it matter? . . . Max McGee wasn't going
to play on Super Bowl Sunday, anyways." At most, he slept a
couple of hours. And when he awoke, McGee had "a sit-in-the-
pit-of-your-soul hangover." Even riding the bench was going to
be rough.

And then, on the very first drive of the very first Super Bowl,
the impossible—or, at least, the sickeningly improbable—happened: Boyd Dowler, the man who'd been starting ahead of him
all season, went down after aggravating a shoulder injury on the
Packers' third play of the game. Max hadn't even brought his

helmet. When Vince Lombardi told him to get into the game, he had to borrow someone else's.

Writes Joshi: "If Max McGee did indeed in that moment feel any trepidation it was surely felt for just a moment and quickly replaced by the aching residue of mass amounts of liquor. But there are reasons some men succeed. There is a reason some become legends . . ."

Somehow—amid the waves of nausea, his dehydrated brain shrinking, his inflamed head swelling against someone else's helmet—the adrenaline pushed all the way through, and Max McGee, bless his hungover heart, broke open down the field late in the first quarter and caught—one-handed—the first-ever Super Bowl's first-ever touchdown pass.

By the end of the game, he had seven receptions for 138 yards and two touchdowns, leading Green Bay to a 35–10 victory. But he didn't win the MVP award; that went to Bart Starr. As Joshi puts it: "The quarterbacks always get the glory. But he did gain something else. That was the day that Max McGee became a legend."

True, without the hangover he might have played just as well, or even better—and may still have made history. But *legend*? That requires a narrative full of obstacles and the most daunting odds—even and especially if you're partially to blame for them. And a legendary hangover is one where you don't mind if it lasts forever, at least in the halls of fame.

THE HUNGOVER GAMES
(THE MASTER AND MALTED WHISKY)

In some ways, Brian Kinsman is the opposite of Max McGee. But he, too, has the makings of a Scottish superhero. Mild-mannered and thoughtful as they come, his well-groomed ginger hair and freckles are as subtle as his lilting brogue. He is soft-spoken, patient and humble: a bagpiper who became a chemist and is now the most prolific malt master in the world.

Every ounce of Glenfiddich bottled in the past half decade (and that's more than any other single-malt on Earth) has his signature on the label. The Malt Master's Edition, found on the top shelf in bars all over the world, even bears his likeness—a photograph of Kinsman, hands in pockets, standing between glowing casks of whisky. And yet, if you were sitting in one of those bars, drinking the stuff right next to him, you'd probably never know it. He is as polite and unobtrusive a man as you'd ever want to meet. The one or two times he's let slip his true identity, in moments of goodwill to an inquisitive pub patron, he's had to leave soon thereafter—made self-conscious by the ensuing camaraderie of every drinker in the place.

But tonight, the man is trapped, seated next to easily the most ignorant, intrusive loudmouth in Glenfiddich's magnificent dining hall.

"So then," I say, slurping on a rare twenty-year-old, "if it's only the barrels that give Scotch its color, why not make a batch of clear Glenfiddich? With the right kind of marketing, I'm sure you could sell the stuff."

Yes, that is what I—who have been drinking whisky since before it was legal for me to do so, and have spent the last couple days being shown and explained every corner of this archaic yet glimmering, visionary operation, and whose job it is to know things, at least a little—have just asked Brian Kinsman, superhero malt master.

But instead of saying, "Well, you know it's not just the color—it's most of the flavor that comes from the wood,"—or "Why exactly were you seated here?" or "Would you please stop talking?"—one of the most respected, and certainly the most gracious, malt masters in the world says this: "Hmm. I hadn't thought of that. So if you see Glenfiddich Clear in a liquor store one day, I guess we'll owe you some royalties."

It is a funny and kind response to an immensely stupid question. But there is also something else in what I'm trying to ask; I have been reading a lot about congeners lately, and the more I learn, the less I know. As far as I can tell, *congener* is a relatively recent, made-up word for pretty much everything that finds its way into an alcoholic drink, other than alcohol. As such, it is congeners that are responsible for the taste and color of a spirit; but they're also referred to as "impurities" and tend to be used as a catchall scapegoat for the severity of a resulting hangover.

As Barbara Holland put it, "These are the tasty additives and enhancements that make our drinks different from one another. They're easily identifiable to the naked eye: darker is bad, lighter is good. Brandy is most lethal, congener-wise, followed by red wine, rum, whiskey, white wine, gin, and vodka. Bourbon will make your head ache roughly twice as fiercely as vodka."

Although it pains me to doubt Ms. Holland, it's an equation I find dubious: perhaps sometimes helpful as a vague rule of thumb, but surely overly simplistic and often downright wrong. Kinsman is more diplomatic: "Well, I think that correlation probably makes some sense in regards to well-made vodka, which is the most highly rectified drink," he says. "Ultimately, you've distilled it over and over so that really all you have is ethanol, maybe sometimes a little bit of methanol, and that's about it. So it's pretty much pure, clear and tasteless. But that doesn't mean those things are dependent on each other. Look at the way that whisky is made."

All hard liquor produced and sold on the world market today is made essentially in the same manner: a grain or root vegetable is ground up, mixed with water and heated, then yeast is added. Fermentation happens over the next few days. Then the residue, often called beer, is boiled over and over again until what you are left with is a condensed, high-proof form of alcohol. This process is called distillation.

Scotch is made from malted barley, often dried over peat fires. That contributes part of the taste and, if you're after it, a peaty single malt. So the smoke residue, for lack of a better term, can be considered a primary congener. Then comes brewing, fermenting and distilling. And when you distill beer, you don't just get ethanol; you get methanol, propanol, butanol, hexanal and all sorts of other long-chain alcohols whose names Kinsman can rattle off like machine-gun fire. Then those combine with the long-chain acids.

"There are hundreds of potential compounds," says Kinsman. "At the low end, they are oils, and at the high end, they are aromatics." Created organically by the process of distillation, these are known as flavor compounds, and they exist everywhere in all we eat and drink. But in pursuit of a good Scotch, they are controlled and balanced at every turn—by the yeast, the temperature, the rate and hours of fermentation and distillation—so that the result is a uniquely defined spirit that is perfectly recreatable, time after time after time.

As much as this is science, it is also magic—starting with yeast, an essential, yet almost unknowable entity. As Adam Rogers puts it, "The miracle of yeast is awesome enough to strain credulity. It's a fungus, a naturally occurring nanotechnological machine that converts sugar to the alcohol we drink. . . . It lives as a single-celled organism that is neither plant nor animal, neither bacteria nor virus."

This fundamental property, without which we would have almost nothing—not bread nor beer, at the very least—grows pretty much everywhere. And although its existence was completely unknown until the mid-nineteenth century, now much of our knowledge of how life works comes from the study of yeast.

It was, in fact, the first creature with both cells and nuclei whose genome was ever sequenced.

Unlike fermentation, distillation was invented by humans— alchemists searching for the essence of life. They developed a process whereby wine or beer was boiled at high heat, with the evaporated alcohol collected in an increasingly purer, more condensed form. Then they decided to store and transport it in oak barrels. But first, they set fire to the insides of the barrels, to remove any lingering taste. And then they noticed something strange: the longer, more tumultuous the journey, the smoother and more delicious the alcohol became.

"So much about whisky is a magic coincidence," says Kinsman. "And no part more than the oak." That same old tree that turned animal skin into leather also transformed rough white lightning into sensuous shooting stars—a delicious comet to the brain that is, quite magically, not so bad for the head. Because here is the thing: if you burn the inside of a barrel, you get two basic layers of scorching—the toast layer and the char layer. When I ask Kinsman the difference, he says, ever patiently, "It's the difference between toasting something and, well . . ."

"Charring it. Sorry."

The toast layer is additive. Kinsman describes it as a kind of tea bag, releasing all these wonderfully desirable things in the wood—color and flavor, soft sugar compounds. These are desirable congeners. "By changing the toast layer, you can influence the eventual flavor."

The char layer, meanwhile, is subtractive. Kinsman relates it to Velcro, "with microscopic hooks that grab onto a lot of the unwanted congeners: heavy sulfurs and some of the more acidic, intense, volatile compounds." So in scorching the inside of an oak barrel, a system is created whereby the spirit is made purer and more complex at the same time—both cleaner and more flavorful.

"If you handle it with care," says Kinsman, "the cask will give and give until finally it's exhausted, and then there's nothing left." To use his mixed yet apt metaphor, the Velcro hooks get worn away and the tea bag runs empty. "If that happens, you'll open up a batch that's been aging ten years and it still tastes fresh, which is not what you want from your whisky. And it might just give you a headache."

"A headache?"

"Could do . . ." says Kinsman, as if just now remembering my weird preoccupation with such things. He holds his glass of Glenfiddich Solera up to the candle, letting the light glance through the amber liquid. "The color can tell you a lot. But it can also lie. If the hue comes from the wood, then a darker color might suggest a more fully aged and filtered spirit—reversing that light-to-dark equation of purity. But if it comes from artificial additives, like caramel, then it could be steering you wrong. There's times I've been . . . drinking young whisky and felt pretty rotten the next day. Whereas sometimes you drink an incredible amount—which, of course, you shouldn't brag about, but you drink an awful, awful lot of really good-quality whisky and wake up feeling fine."

It does seem there's something to this. After all, I can't usually—or really *ever*—afford to drink so much top-shelf stuff, and since I arrived at Speyside, the mornings have been relatively painless. What a beautiful, impossible alchemy of ancient aesthetic, ascetic and anarchic that would be—whereby the sweetest yet most subtle, even accidental, keys to pleasurable scent and taste might also save our heads.

"Well, I suppose our senses work that way," agrees Kinsman, looking at the liquid in his glass. "We're warned off things that can harm us and drawn to things that don't."

"Knock on wood," I say, cringing at the pun. We clink our

glasses together before taking another sip, and I'm reminded of why we do that—or, at least, why people think we do. One explanation is that, long ago, when folk sat down for a drink, they'd bash them together first, letting the liquid slosh from one to the other—safeguarding, in a somewhat friendly way, against the worry of your cup being poisoned.

Another, more poetic, idea is that when you come to a drink, it awakens four of your senses: the sight of it in your glass, the touch of it to your lips and, of course, the scent and taste. Then, by clinking your glasses together, you hit that fifth sense—sometimes with a perfect ringing note.

"Do you know what I like most about it all?" says Kinsman.

I tell him I have no idea.

"It's the way a spirit keeps on changing, even once it's in your glass. We work so hard to make a consistent product, and yet it will always taste different, depending on infinite things: your mood, what you ate, the air, the company. You suddenly find something you never would have guessed."

For anyone else, there might be something fanciful to this. But it is this man's job to be in complete and full awareness of his senses. He creates color, texture, smell and taste. He takes a sip of his own creation and describes to millions of people what is in their glass.

And yet, what he's really saying, as the maker of the most popular single-malt in the world—a product renowned for its remarkable consistency—is that what he likes most of all is its unpredictability: as amorphous and subjective in a single sip as whatever might happen if you take a hundred more. "How about you?" he says, pointing to my glass of Solera. "What are you getting from it?"

I take a whiff, and another sip. "It's good," I say. Trying harder, I come up with: "Honey, orange, violins . . ." But then I think I might just remember *honey* from Glenfiddich's sampling notes, which of course Kinsman wrote. I'm not even sure if by *orange*, I mean the taste or the color; the room is aglow with orange flickering candles. And now I notice that a woman in the corner is playing the violin—so maybe, rather than sensory synesthesia, what I've got is a woefully unrefined brain, with little ability to really access, let alone understand, what's going on inside it.

Kinsman's brain, on the other hand, through years of concentrated sampling, has created neural pathways by the tens of thousands that reach into every part—allowing him to recognize the characteristics of a single vapor instantaneously, wordlessly, but also find something new, which he tries to put into language.

He lifts the glass to his nose and thinks something like, *Sweet . . . waxy—the kind of sweet and waxy when you open a bag of raisins, a particular kind of candy too—one I first had when I was six, at my grandmother's.* And now something he's never noticed, though he's been both drinking and making Solera for years—like a magician surprised by his own magic: *Licorice, with smoke, with . . . steam; the smell of rain on hot pavement . . .*

Kinsman smiles. "You're right," he says. "It's good."

SCOTLAND'S *OTHER* NATIONAL DRINK

The actor, comedian, musician and modern poet laureate of Scottish hangovers, Sir Billy Connolly, has a rousing lament entitled "The Afternoon after the Morning after the Night Before." On a 1974 concert recording, he introduces it like this: "I'd like to dedicate this wee poem to the makers of Irn-Bru, Mr. and Mrs. Barr and the wee ones, for saving my life on so many Sunday mornings."

As obscure as that sentence might be to everyone else, in Scotland—the only country in the world where a local soda pop outsells Coca-Cola—the legacy of Irn-Bru is up there with Braveheart. *The Oxford Companion to Food* defines it as "a Scottish soft drink, which is important for its symbolic value as well as for its refreshing qualities." Irn-Bru is more than a soft drink, more than a symbol. To the Scots, it is, most mythically, their non-alcoholic, sweetly ironic, hard-as-nails hangover cure.

First bottled in 1901 by Robert Barr and his son Andrew Greig Barr (the company is named A.G. Barr), it has been around almost as long as Glenfiddich, with which it shares the distinction of still being a family business. The stuff used to be called Iron Brew, but that was changed in the '40s since it isn't actually brewed and the question of how much iron it contains touched too close to the mystery of its secret ingredients.

According to *A Dictionary of Scottish Phrase and Fable*: "The recipe, which is kept in a bank vault, is known only to two people: the chairman—who mixes the ingredients once a month in a sealed room—and one other unnamed employee. The two are never allowed to fly on the same plane together."

For an actual Scotsman, the lore of Irn-Bru is far more important than that of the Loch Ness Monster. In the 1980s, the Barr family came up with the slogan "Made in Scotland, from girders"—suggesting untold strength, along with a nod to the iron-workers of Glasgow, but hardly revealing of its actual ingredients. And in fact, the iron gates around Irn-Bru have proven very tough to breach.

Failing on my own, I asked the good folks at Glenfiddich—whose former accountant now works at Irn-Bru—to try to arrange an introduction. Failing that, I pulled out the big guns and turned to Jonathan Dart. Among his many international missions, our man Dart also happens to be involved in brokering a deal to get Irn-Bru into the Canadian market. But even he, so far, has only found new gates for me to bang on.

Much like Red Bull, the Austrian energy drink, Irn-Bru has become a prolific mixer in the bars, clubs, homes and pubs of Scotland and beyond. But the stuff was never meant to be mixed with booze. Even this whole hangover business is pretty much the opposite of its original intent, which was to beckon folks away from the ills of alcohol. According to *The Oxford Companion to Food*, "The impetus towards commercial production of such drinks lay partly in the strength of the temperance movement." Which may explain why no one's responding to my banging.

The temperance movement, which would eventually affect all of Western society in very odd ways and lead to Prohibition in America, was first fostered in Europe by a dismal trinity: the Gin Craze, the Industrial Revolution and, finally, Cholera.

As opposed to the bubonic plague, for which drinking booze was pretty much the only defense, cholera was seen as being exacerbated, if not directly caused, by heavy drinking. Trend-setting teetotalers pushed this idea, particularly in Scotland, where Calvinism, the Scottish work ethic and growing production of Scotch whisky made for an increasingly uneasy relationship with the bottle.

Looking at a public-service poster issued by the Scottish Health Board in 1832, you'd be justified in thinking that cholera was in fact a deadly form of hangover specific to the underclass. In a somewhat cosmopolitan turn, the health board took the reader on an around-the-world cholera tour (all emphasis is theirs):

RUSSIA.
"It had been observed at Moscow and Riga, that any great Festivals, where the Lower Orders were assembled, and

where <u>intoxication</u> was a common consequence, were always followed by a marked increase in ensuing Day's List of Invalids. . . .

"The effects of previous <u>intemperance</u> on the system seemed to predispose it more than any other cause to the Disease."

—Lefevre, on the Cholera at St. Petersburg.

INDIA.

"Persons of sober, regular habits, enjoyed greater immunity than the <u>drunken</u> and <u>dissipated</u>, who kept irregular hours, and were frequently exposed to the vapours and cold of the night after a debauch."

—Jamieson's Report on the Cholera in Bengal.

POLAND.

"Three Warsaw Butchers went to a Tavern, abandoned themselves to every sort of excess, and drank till they were so intoxicated that they were carried home senseless. A few hours had scarcely elapsed, when the miserable men were seized with all the symptoms of Cholera, which advanced with such rapidity <u>as to prove fatal to the whole Three within Four Hours</u>."

—Brierre de Boismont, on the Cholera in Poland.

GERMANY.

"The great majority of persons attacked with Cholera in Berlin consists of those who are exposed to the usual causes of Disease—namely, cold, fatigue, and <u>particularly intemperance in food and drink</u>."

—Dr. Becker's Report on the Cholera in Berlin.

Eventually, the temperance movement in Scotland won a single day of the week. The Licensing Act of 1853 prohibited drinking on Sunday, except by "bona fide travelers." This led to locals traveling to neighboring towns to get pissed on the day of the Lord, and a whole new stranger-in-a-strange-land aspect to Monday hangovers.

But the most ardent mobilization of the temperance movement, at least in terms of propaganda, was focused on the kids—most famously by George Cruikshank. A once-heavy drinker known for illustrating the novels of his friend Charles Dickens, Cruikshank became a sort of artist-in-residence for, and then vice-president of, the National Temperance League. He produced dozens of texts and engravings aimed at inspiring future generations to never pick up the bottle—in fact, to smash it.

In *The Bottle*—a series of eight engravings—we see the patriarch of a loving family taking a drink for the very first time. Then, just seven plates later, "The Bottle has done its work—it has destroyed the infant and the mother. It has brought the son and the daughter to vice and to the streets, and has left the father a hopeless maniac."

But there was still more work to do. In *The Drunkard's Children*, an eight-plate sequel to *The Bottle*, we catch up with the son and daughter drinking in a gin shop. By the final scene, "The maniac father and the convict brother are gone. The poor girl, homeless, friendless, deserted, destitute, and gin-mad, commits self-murder."

These family-centric fables were apparently inspired by Cruikshank's own father, an abusive drunk. Cruikshank's past days of boozing, and the accompanying hangovers, provided enough material for his depictions of the more immediate and physical repercussions of heavy drinking. His 1835 engraving *The Head Ache* speaks painfully for itself:

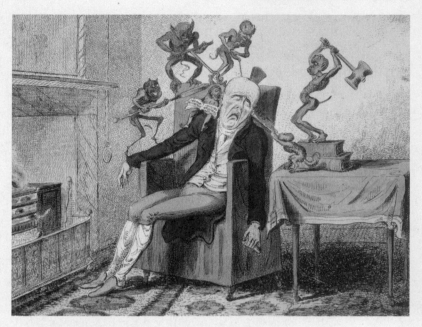

THE HUNGOVER GAMES (THE LAST DROP)

It is my final evening in Scotland, and I'm sitting with John McFetrick, drinking beer and whisky at the Last Drop. It is a great name for a pub, and here it is practically a triple entendre. They once hanged deviants in the street outside, and the wooden sign above the door depicts a swinging noose. But also it was places like this one, Edinburgh Castle looming above it, that helped coin the expression "to go on the wagon."

Back in the day, condemned prisoners were taken from whatever dungeon or tower they'd been locked in, put on a horse-drawn cart and paraded through the streets to the public gallows. Sometimes, if the magistrate was feeling magnanimous or the condemned prisoner was well liked, they would stop at the pub for one last drink. The prisoner would empty his glass, presumably very slowly, and then it was time to get "back on the wagon."

"Well, that's both informative *and* cheery," says McFetrick, in that sardonic Eeyore tone that hasn't changed since university.

At the time, though we adored each other, we were not very similar. But now, twenty years later, we have a lot in common: two mostly Irish Canadians drinking in Scotland, we both have young sons, complicated domestic situations, profound contradictions embedded in our souls, a love for whisky, strange demons lurking around us and a dangerous habit of messing with them.

McFetrick, for example, has spent the last long while studying thousands of hours of videotaped conversations with manic-depressives institutionalized at the University of Edinburgh, while I have been studying hangovers. We knock our glasses together: to old times and new ironies.

We order some haggis and I ask him about his hangovers. (I recall him having some doozies.) He tells me about one before I knew him, when—in sweltering heat, presenting his squadron while the bugles were blowing—he somehow puked on his drill sergeant without the man ever noticing. He laughs about it—and then about the time I got us busted by the Montreal transit police.

It was Valentine's Day, and he and my longtime girlfriend Ibi and I decided we'd all eat hallucinogenic mushrooms, drink a lot and go to a late-night screening of *Caligula*—as you do. Then, on the way there, we convinced McFetrick to jump the turnstile. He tripped, the subway cops descended, and a very awkward sort of chaos ensued.

"Those," says McFetrick, "were the good old days."

Looking around, it is easy to imagine Robert Louis Stevenson and Arthur Conan Doyle tying one on in this very bar. They were, after all, drinking buddies and classmates at the university here.

Traveling these weeks around the UK, I've noticed several connections about the route I've taken. I show McFetrick the poker I made at the blacksmith's forge in that little Devon town, which it turns out is right next to the Buckfast Abbey.

"The Buckfast Abbey," I say, as our haggis arrives. "Do you know what that is?"

"A place to put monks?" says McFetrick.

"Well, yeah. But it's also where Buckie comes from."

BUCKFAST TONIC WINE—OR Buckie, as it's known on the streets of Scottish cities—is the worst nightmare of both the Scottish police and people like Malt Master Kinsman. It is ridiculously cheap and contains 15 percent alcohol by volume, infinite dubious congeners, tons of refined sugar and more caffeine than eight cans of Coke. Simply put, it will mess you up. And apparently, it is the drink of choice for Neds, a term that stands for "non-educated delinquents."

According to Scotland's largest police force, Buckie has been referenced in more than five thousand crime reports, with Buckfast bottles themselves used as a weapon in more than a hundred violent assaults. So the authorities called out both the punks and the monks, and local police started marking bottles to trace their purchase.

In December 2013, Scottish health secretary Alex Neil called for the brothers of Buckfast Abbey to stop production. But the English monks, as they're wont to do, remained silent. They just kept making that good ol', bad ol' Buckie—and sending it all to Scotland. And so now, the bottled-up monster, the transformative tonic, the vibrating fear of a beast within, has come back in new form to these Scottish cities beneath the foothills of the Highlands.

We drink Scotch and beer until the pubs are closing. I hug my old friend in the street and tell him I'm going to hail a cab to my hotel. But even as I say it, my mind is changing, until my actual plan is to find some Buckie—just to see what the stuff can do.

A HELLUVA WAY TO WAKE UP

On top of a bus shelter in Berlin.
—Pat Fairbarn, 33. Smart hipster. Works in a cheese shop.

In an empty chemical lab in Tampa.
—Ken Murray, 44. Writer. Teacher. Rower. Appears elsewhere in this book.

On a rickety old sailboat in a super-fancy marina in Mazatlán, fully clothed but with two naked men I barely remember passed out on the floor. I had just enough time to open my eyes and gather my things before they set sail to the Galapagos.
—Lindsey Reddin, 31. Massage therapist. Adventurer.

In my rooming house on the floor surrounded by firemen in full gear, massive boots just missing my head and hoses crisscrossing the floor. I crashed out with a lit cigarette and my friend wakes up to find a dinner-plate-size smouldering hole beside my face and pours a pot of water on it, rolls me off, and tosses the mattress onto the wooden balcony where it keeps smouldering until one of the neighbors calls the fire department. Two things learnt: mattress fires are almost impossible to put out, and firemen have absolutely massive boots.
—Bruce LeFevre, 58. Fellow writer. Fellow drinker.

In my bed . . . woke up with a half carton of melted Häagen-Dazs. Thought it was something else entirely.
—Melinda Vancuren, ER nurse. A great person to drink and eat ice cream with.

*Behind a rack of barrels in the middle of a Niagara winery
tour. Sue will confirm.*
—Robert Hough, 51. Brilliant author. Dubious storyteller. Lucky
husband of Sue.

In a kitchen I'd never been in before and I had no pants.
—Duncan Shields, 41. A high school friend. Was very smart and tall.
Probably still is.

*The answer, SBS, is "on your kitchen floor with my face
in a jar of dill pickles."*
—John McFetrick, 48. Always did like pickles.

THE HUNGOVER GAMES (CLOSING CEREMONIES)

I awake to a rattling and buzzing in my weirdly hot skull. I am
half-clothed and sideways in this room that barely fits the bed.
My feet are in the air and my head is on the heater beneath the
window, from which a stunning view of Edinburgh Castle is
mostly obstructed by a card: "Guest Notice: Please do not cover
or obstruct any of the heat outlet grilles or the air intake open-
ings of the heater."

I attempt to unobstruct the grilles by somewhat raising my
head, but it feels like I'm trying to pull out my brain. So I reassess
my options, then maintain my position while moaning softly.

An hour or so later and I'm sort of up, glugging water
from the miniature bathroom at the foot of the bed. I pull on a
shirt, check out of the hotel and drag my bags to the bookstore
next door. The headache has lessened, but there's still a heavy,
downward-driving queasiness.

It is a fancy, modern sort of bookstore, with a café on the
second floor, among the cookbooks, self-help and Penguin

classics. And right there, beside a display case of croissants on
the café counter, is a mini-fridge of Irn-Bru. I buy a can and sit
at a table by the windows, with an unobstructed view of the
hilltop castle.

The aluminum is nicely chilled, contracting slightly as I crack
the top, then take a sip. The taste is much better than I thought
it would be—sweet, but cut right through with a tangy kind of
steeliness: like cold orange irony. As I drink it, I start to feel better. And now it's time to sort my head out a bit—put aside bestial
thoughts and get back above the ground, onto a plane and home
to Canada. I stop at a gift shop to buy presents for my boy and
my girlfriend.

I choose sexy Tartan tights for Laura and a teddy bear dressed
as Darth Vader for Zev. Then I tuck them into my suitcase, along
with a six-pack of Irn-Bru, and head toward the airport.

A ROOTS OF REMEDY ROUNDUP

Over the past little while, people keep sending me articles about South Korea's ice cream hangover cure. This is what happens when you've been working on something for too long and everyone knows it. Last year, my social media accounts were full of Chinese researchers who—speaking of Irn-Bru—had recently decided that orange soda pops are the perfect thing for the morning after. Before that, it was Japanese pears, carbonated cabbage juice, fermented tomatoes and a hundred ways of cooking eggs.

But tomatoes, cabbages, pears, sodas and breakfast have all been around for a while now. And the exciting new ingredient in that South Korean ice cream comes from a raisin tree that's been used in hangover remedies since the 1600s.

I'm not saying that some magic-bullet cure couldn't possibly be lurking within things we already know—only that we already know them. So here, for the purpose of continuity, historical interest and some insight into what we're working with, is a quick roundup of hangover remedy ingredients that have stood the test of time—by which I don't necessarily mean they've been successful, but that people have kept on trying them, in different ways, for hundreds of years, and sometimes still do.

Pickled herring, for example. As a remedy, it dates back to when we started pickling fish in the first place. The Germans even have an on old herring-based hangover euphemism: "That man is in need of a herring." And cheeky university students in thirteenth-century France played something called "the herring game" when filing into Mass on Sunday mornings. It involved tying herrings to their gowns so as to trail on the floor behind them. The object of the game was to step on the herring in front of you while somehow protecting your own. But the real purpose of the herring game was to torment the pious clerics with thoughts of your heathen drunkenness the night before. To this day, the breakfast buffets of Europe, Scandinavia, Russia and beyond are lined with fishes whose overwhelming oiliness is thought to neutralize alcohol's acidic effects on the stomach.

Even more prolific as a hangover remedy is the most common northern side dish to an oily fish: sauerkraut—or, at least, its main ingredient. All the way back to the ancients, people have used cabbage—pickled, cured, raw and, most commonly, boiled—as both preventative and cure. Wrote the Greek rhetorician Athenaeus, in the third century AD: "That the Egyptians are wine-bibbers is indicated also by the custom, found only among them, of putting boiled cabbage first on their bill of fare at banquets, and it is so served to this day."

Apparently, Aristotle even had a pithy little rhyme about it:

Last evening you were drinking deep,
So now your head aches, go to sleep;
Take some boiled cabbage when you wake,
And there's an end of your headache.

If any of this is true, it is probably due to cabbage's chelating abilities—meaning it can bind with toxic elements in the body and carry them out as it leaves. Other natural chelators include milk thistle, guava leaf, cilantro, charcoal and N-acetylcysteine— all of which are active ingredients in recently marketed hang- over products. In fact, the modern era of manufactured remedies pretty much began with Chaser, a pill that was essentially just charcoal—or "activated carbon." Though now discontinued, it is widely considered, at least from a business perspective, one of only two successful hangover products to date, the other being NoHo—a liquid shot, apparently created by a New Orleans doc- tor, in a plastic bottle with prickly pear extract as its main active ingredient.

Also known as the nopal cactus and Indian fig opuntia, prickly pear contains betalains—a rare, colorful class of antioxidants that give beets (another historic hangover remedy) their almost unearthly redness. Prickly pear has been used as sustenance and medicine, at least in Mexico, for hundreds of years. Then, in the early 2000s, the entrepreneurs behind NoHo sponsored a double-blind, placebo-controlled, clinical hangover trial with encouraging results . . . sort of.

Although it was found that prickly pear extract "did not reduce overall hangover symptoms," it did "reduce the risk of severe hangover by 50%." The subjects "reported nausea and dry mouth less frequently," but symptoms of "headache, sore- ness, weakness, shakiness, diarrhea, and dizziness were similar to the placebo group."

Researchers attributed the benefits not only to prickly pear's antioxidant qualities, but to its strong anti-inflammatory effects— supporting the theory of hangovers being, at least in part, an inflammatory response. Although no further research has been published, that one study was apparently sufficient for NoHo's marketing team, allowing them to put the words CLINICALLY PROVEN TO WORK! on the packaging.

The most recent new-and-ancient remedy "clinically proven to work" is the Japanese raisin tree, aka *Hovenia dulcis*, source of dihydromyricetin, aka the stuff in the Korean ice cream bar that's making all the headlines. And how it has become known (and known as so many things) says a lot about the hangover remedy industry in general.

In *Proof*, Adam Rogers takes us through the rediscovery of Hovenia via Richard Olsen, a neuroscientist at UCLA. Olsen is of the current school of thought that the way we respond to alcohol (from drunkenness to hangovers to alcoholism) has a lot to do with the neurotransmitter gamma-aminobutyric acid (GABA) and its receptors. Recently, he became interested in one particular "extrasynaptic" GABA receptor that responds in a complicated fashion to concentrations of ethanol as slight as those produced by a single glass of wine. Thinking this might be the key to all sorts of alcohol-related dysfunctions, including hangover, his team began to hunt for a drug that would bind to it in an effectively neutralizing manner.

And so, as Rogers explains, "one of Olsen's postdoctoral students, a researcher named Jing Liang started experimenting with herbs from her native China, beginning with the ones that traditional medicine claimed had an effect on alcohol." And that's when she found bits of the Japanese raisin tree on the shelf of a Chinese grocery store.

Olsen and his team scienced the crap out of it, purified it in a laboratory, isolated an ingredient they called dihydromyricetin, made pills out of the stuff, then passed them around to colleagues who were headed for the bar. Users reported feeling less hungover the morning after—but also more sober the night before.

As far as I can figure it, the company that ended up backing Olsen's research came out very quickly with a product that appears to be a hangover remedy . . . just not in so many words. BluCetin (an imprecise name if ever there was one) is created by the Sundita (whole health and vitality) company, whose web page asks and answers questions like this: "Tired of that awful, toxic, rundown feeling the morning after a night out with friends? . . . Try BluCetin today and help provide your liver and body with the added support it needs to keep up with the demands of modern life."

Then, soon after, other Japanese raisin–based products started to appear, but without so many euphemisms. Sobur, for example, is much more direct about what to expect: "Dihydromyricetin (DHM) will reduce the degree of drunkenness for the amount of alcohol drunk and will definitely reduce the hangover symptoms."

By the time I get in touch with Dr. Olsen to talk about this, he seems altogether chagrined, resolute and resigned. "I would like to state," he states, "that we (Jing Liang and I) don't really believe that DHM is a great hangover cure, more so a blocker of alcohol action and thus intoxication and whatever results from that."

To Olsen's credit, what he is helping identify here is a continuum relating to the age-old, still somewhat baffling linguistic, societal and even scientific confusion between the problems of drunkenness and those of hangover. As for my own pursuits, I am starting to move purposefully past those remedies that, if they work, do so by first remedying drunkenness—as that does seem to defy, at least in spirit, the purpose of our quest.

As for these various ingredients and products, I have spent the last long while boiling cabbages, scraping charcoal, steeping Hovenia, dissolving powdered prickly pear, popping pills and eating oily fish—before, during and after drinking all kinds of different booze. It can be messy, stinky, dodgy work, and often the resulting hangover seems worse than if I'd tried nothing at all. But then, other times—and more and more lately—I wake up clear as a bell, right as rain, feeling no pain, and I get downright optimistic.

PART SEVEN

THE FUTURE'S SO BRIGHT

IN WHICH OUR MAN ABOUT TOWN THROWS SOME PARTIES,
THEN DINES WITH A LEGENDARY CLUB OWNER IN AMSTERDAM.
APPEARANCES BY HIERONYMUS BOSCH, A BAR FULL OF
MONKEYS AND A MAN NAMED CHEESEBURGER.

*On some days my head is filled with such wild and original
thoughts that I can barely utter a word. On other days,
the liquor store is closed.*
—FRANK VERANO

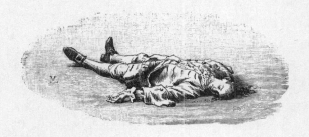

T HIS IS A GOOD LITTLE PARTY: A DOZEN PEOPLE and a half dozen bottles of Irish whisky in this classy, rugged Vancouver loft owned by a classy, rugged cinematographer. He and his writer friend Masa heard about the research parties I've been throwing and were kind enough to invite me to this one—their first annual St. Patrick's Day whiskey tasting—provided I bring some remedies.

Over the past few months, I've been testing all sorts of combinations of pills, powders, tinctures, teas, emulsions and energy drinks—first on myself, then on friends and family, at poker games, weddings, wine tours and now here, at this good little party of trusting strangers ranging in age from twenty-three to fifty-six, including an architect, an ecologist, a gardener, a mental health worker, a medical educator, a video-game designer and a half dozen thirsty writers.

To get their attention, I clink on a glass of Tullamore Dew, thank them for offering their livers to science, then run through the rules. Each of them will receive one of nine different kinds of concoctions, along with instructions on how and when to ingest it. These include two manufactured hangover products, one powder to be dissolved in water, some herbal capsules, a liquid mix provided by Bronwyn of Gaia Garden, an Ayurvedic tincture created by Todd Caldecott and a mixture of pills I've been working on.

The nine possible remedies contain, among them, over forty separate ingredients—including milk thistle, chicory seed, barley grass, burdock root, licorice, Chinese foxglove, dandelion, ginseng, ginkgo, turmeric, kudzu, calcium, folic acid, a number of vitamins, magnesiums and amino acids, chicory extract, grape extract, palm extract, prickly pear, caper bush, arjuna bark, yarrow flower and black nightshade, as well as some other obscurities. Some people will be ingesting only one of these ingredients, others more than a dozen. But none will know what they took until tomorrow's debriefing—or the following day's, should it be that long before they start to feel better.

I thank them for their trust and request that they eat some of the food provided, mark their drinks on the chart taped to the closet door and allow me to Breathalyze them at least once during the proceedings—other than that, the night is theirs. I wish them a happy St. Paddy's and we raise our glasses.

Five hours later, the classy, rugged cinematographer is vomiting in his own bathroom; a husband and wife have left after getting into an argument; the twenty-three-year-old writer has taken charge of the music selection with all the subtlety of a mob boss on psychotropics; the ecologist is giving a lecture on wine; the mental health worker is dancing with the gardener; and I am still testing the whiskies.

FLY BY NIGHT

At any one time, you can find about eighty available products worldwide that claim to mitigate, alleviate, remedy or cure hangovers. I have reached out to dozens of such companies, and even been in touch with their founders. But what keeps on happening is that, once I've got things together regarding travel and interview plans, they're already packing up shop.

Now, admittedly, this might have something to do with the elaborate scope of my imagined research schedule and the increasingly haphazard pace at which I appear to be working, but it also surely says something about this apparently burgeoning, possibly collapsing, oversaturated, inexplicable industry I'm trying to learn about.

The brutal truth is that more than one originally optimistic hangover remedy entrepreneur, after just a few correspondences over a matter of months, has then suggested that I—a writer who tries to supplement the income of his books with journalism, his articles with teaching, his courses with writing—could be the perfect person to *purchase* their hangover remedy company.

But as foolish as it all seems, I'm not sure that any of these companies are failing because their products are not good. Certainly, they might suck, but more to the point could be their approach—or lack thereof—and the general temperance, state, faith and imagination of the people on this planet.

And I'm starting to suspect that confronting hangovers, let alone solving them, pertains to much more fundamental things than we've yet begun to consider.

THE FUTURE'S SO BRIGHT (FOR WRONG OR FOR RIGHT)

Looked at from afar, the St. Paddy's Day party was a great success. I have garnered charts, notes, anecdotes, Breathalyzer results and a number of positive testimonials. In their own words, here are a few.

From Adrienne Mathews, age twenty-three, after five beers, five whiskies and a BAC of 0.12. Remedy: powdered *Opuntia ficus-indica* (nopal/prickly pear):

> *I felt pretty good all night, just very voluble and over-share-y. At some point I lost my tolerance for all music*

*I had not chosen personally, even though retrospectively
I wasn't being super on point with my own DJ choices.
Probably went to sleep around 5 a.m. and woke up at
nine, had the shakes and felt like total baloney, but by
3 p.m. I was out for a walk, half being like "I am sub-
suming the calm and beauty of nature" and half thinking
I might barf, which I didn't.*

From Robin Esrock after two beers, four whiskies and a BAC
of 0.06. Remedy: kudzu root:

*I expected to feel alot worse on Sunday morning, mostly
because it's been a long time since I've gone to bed tipsy at
2 a.m., with a teething two-year-old up all night ensuring
any quality sleep would not be forthcoming. My daughter
was up up at 6:30 a.m., my wife was just fed up and so
I had to get up up as well. All recipe for a lousy morning
hangover, but there wasn't one. I credit this to: (a) the
placebo effect of whatever the hell was in those pills, (b)
whatever the hell was in those pills, (c) the fact I didn't
come close to Jeff's impressive tally on the scoreboard.*

From our host, Jeff Topham, age forty-three:

*By official tally, I had 6 bottles of beer and 9 whiskey-
based drinks between 7 p.m. and 3 a.m. (Considering I
don't remember much after 11 p.m. and may have neg-
lected my record-keeping, the count may well have been
higher.) At the end of the night my blood alcohol was
0.13. Around midnight (approx.) I discovered my eyes
weren't working properly and my stomach needed to
reject its contents (i.e., vomitus). After a brief (15 min.*

*approx.) repose/paralysis on my bed upstairs, I returned
to the festivities to continue drinking. Many would see
this rally as "heroic." As instructed, I had taken my
"remedy" (#3) consisting of three (3) small pills prior to
the onset of imbibation, and one large amber pill shortly
before bedtime.*

*Normally, such an evening would render me incapaci-
tated (i.e., "fucked") . . .*

*But today is unlike any day-after-the-night-before
I've experienced in 25+ years of chronic booze artistry.
I awoke at 9:30 a.m., still a bit fatigued but otherwise
miraculously unscathed. My head didn't hurt. My mind
was uncannily clear. I actually texted "It's a miracle"
to a couple friends who were inquiring as to how the
experiment had gone. The remarkable fact is that it
was a rather unremarkable Sunday. I did some work,
watched some TV. My body felt tired, but I definitely
did not feel like I had consumed likely over 15 drinks on
a mostly empty stomach.*

*I swear the above account to be true and hyperbole
free. Thanks for including me in your worthwhile project. Please let me know if you have any other follow-up
questions, and also what the fuck was in that magic
amber pill?*

That magic amber pill was full of good old frankincense, and
remedy number 3 is the concoction of my own making, with
N-acetylcysteine as the main ingredient. No scientist would take
such pride in such a woeful experiment, but I allow myself to
think I may be on to something, if only for myself.

The test evening had its downside. And brings us to Randy
Baker.

In a number of ways, he is a singular test subject. He was the only one drinking red wine instead of beer and whiskey, and also the only tester of remedy number 11—a herbal product containing caper bush, chicory seed, black nightshade, arjuna bark, yarrow flower and tamarisk plant. I'm not really sure what any of those things are, but it's safe to say I won't be using them again. Here's what Mr. Baker had to report:

> *I don't think that I have ever been as ill after drinking in my life. . . . Ever. I consumed a bottle of wine, 3 whiskey shots, a meal and several glasses of water from 8 p.m. to 1 a.m. Not an excessive amount. I left Jeff's at 1 a.m. or so, having blown only 0.1 and feeling tipsy but not too bad. . . . During the journey home, first on foot then in a cab because I wasn't sure I could make it, I seemed to get increasingly drunk and disorientated, with the world getting "fuzzier." I made it in the door but I don't remember brushing my teeth or how I got into bed. . . . When I did get into bed and closed my eyes, I got the spins as bad as I have ever experienced them. . . . I stumbled out of bed and lurched to the bathroom where I spent the next 15 mins evacuating my stomach contents, mostly hitting the toilet and/or sink. Feeling a bit better after that, and with nothing left to give, I retreated back to the bedroom and passed out.*
>
> *I came to about 0900 . . . spent the entire day feeling poisoned, crappy, tired and sorry for myself. Even after waking up on Monday a.m. I felt fuzzy and strange. It wasn't until about 1 or 2 p.m. that I seemed to recover enough to realize that my liver hadn't dissolved and that I might actually survive. It was at least a couple of days before I could happily face another glass of wine.*

In relating this experience to a couple of friends and
my girl, Joanne, who watched me suffer thru it, more
than one person said, "so . . . . you went to a party, met
a stranger who encouraged you to get drunk, gave you
some pills that you didn't know what was in them and
then sent you home alone . . . that about sum it up?"

Yes, it does.

Cheers, Randy.

MEA CULPA (AND ANOTHER POSSIBILITY)

Randy, I am sorry. I truly am. No matter how glib I may have
seemed on St. Paddy's Day, that was just my Irish coming out.
Not for a moment was I taking people's trust in me, nor my brief
position of power over them, lightly. Or, if I was, I certainly won't
do it again. I accept full responsibility for whatever that dubious
concoction of bush, seed, bark and nightshade may have done to
you, and I have struck it from the list.

There is, however, another possibility. It is at least worth noting that, while the rest of us were guzzling beer and Irish whiskey,
you alone were drinking red wine.

Whenever I'm asked (and also quite often when I'm not), I say
that if I were allowed only one drink for the rest of my life, it would
be red wine. I could write whole chapters about how good it is
(never mind how supposedly good it is *for* you). But no matter how
much I love the stuff, there are people in my life I love just as much
who can't even touch it. In fact, it seems every day now, someone
else has to give it up entirely because the pain of not drinking red
wine is finally outdone by the pain of drinking it, because of the
hangovers—and even more, because of the goddamn migraines.

In recent years, red wine migraines have topped the list of
first-world problems. But as opposed to cordless earbuds and

poodle angst, this definitely deserves our attention. As tragic as it is, the phenomenon is also mysterious and, I'm starting to suspect, might be indicative of much larger things.

According to the most common of knowledge, these adverse reactions to red wine (sometimes exhibited after only one sip) are the fault of either sulfides or tannins—or possibly both.

Our leather tanner back in England, though endlessly pragmatic, referenced "magic" to explain what tannins are and their place on the skin of everything. Luscious, full-bodied reds are made that way mostly by the tannins on the skins of the grapes languishing as they ferment. And they're the ones causing instant migraines and migraine-type hangovers. But those wines that barely sit with the skins, or touch them not at all—the thin reds, rosés and whites—seem relatively innocuous in this regard. So, clearly, it must be the tannins.

But if that is true, why only now? Why, after hundreds, if not thousands, of years of documented wine production, can I find not a single reference to well-made red wines causing the kind of apparent systemic and even immediate pain they seem to be causing now—not even in my parents' early drinking days? In fact, it wasn't until just a short while ago that I started putting things together—when my dear sister Cassidy, whom I don't get to see nearly enough, invited me over for dinner.

"You don't have to bring anything," she wrote. "But if you do, I can't drink red wine unless it's organic. Otherwise I get migraines. Weird, right? I guess I'm a full-on flake now. But really, you don't have to bring anything."

Flaky or not, my sister is not alone. Over the next while, I started bringing organic wine to all sorts of people who'd sworn off reds because of migraines or debilitating hangovers, and for nearly all of them it did appear to work. But why? Tannin levels should be no different, whether a wine is made organically or not.

So, then, there's the sulfides.

Though a necessary part of the winemaking process, sulfides appear in at least half the usual amount in wines made organically. In fact, they have to for them to be labeled "organic." And it is feasible that—as with tannins—certain people may have an extreme sensitivity to sulfides, causing allergic, asthmatic and migraine-like symptoms.

But there is also another theory, one that's been growing in my mind as I've talked to leather tanners, gut researchers, winemakers, my beloved sister and other good people . . . and I am starting to think it may explain why full-bodied reds appear to be the source of such distress, why it seems a purely modern phenomenon, and also why organic wine, properly produced and carefully consumed, might cause no pain at all.

THE FUTURE'S SO BRIGHT (WITH FEWER RED LIGHTS)

"So, want to know the culprit?" I say, holding up my glass at this table that feels like a bold spotlight on this grand, sweeping, architecturally brilliant stage that is Brasserie Harkema, one of the finest new restaurants in central Amsterdam.

And I am tossing, quite carelessly, this slightly obnoxious rhetorical question (referring to an answer to a question that nobody asked but me) to Michiel Kleiss—who owns this spectacular restaurant in which we're sitting; who is to thank for the sumptuous, carnivorous meal we're devouring; who bought the wine we're now drinking (a rare-but-not-impossible grenache from hundred-year-old vines in the Barossa Valley, which he won at auction and describes as "Châteauneuf-du-Pape on steroids"); and for whom I've traveled to Amsterdam in the hopes of learning all kinds of things.

"Okay," he says, "what is the culprit?"

I nose and sip, then take a glug and put it down.

"Pesticides."

Probably only in my imagination do the diners go quiet and push back their chairs, ready to rise against me, while Kleiss coolly waves them down.

"As I'm sure you know," I continue, "there are few fruits in the world more saturated with pesticides than wine grapes. They're immensely valuable—whole economies depend on protecting them, right? And pretty much all they are is surface area: little spheres of juice wrapped in skin. And of course, that skin is where the enzyme is that allows for fermentation. So they can't be washed before being crushed. To make a white wine, they get rid of the skin—but the red needs to stew with it, longer and longer, and in various ways, depending on the wine's intended robustness.

"So, really, when people attribute the apparently recent phenomenon of red wine migraines and severe hangovers to tannins (which have been a part of it all forever) or sulfides (which are only curbed by 50 percent in most organic wines), might it not be worth looking into the toxic pesticides—of which there are dozens, if not hundreds of kinds—coating each of the thousand grapes in a single bottle of wine?"

Okay, maybe I'm not precisely so eloquent, and the blather doesn't quite end in a rhyming couplet. And perhaps it's not the classiest toast to make with a wine your host has bought at auction. But over the past few months, I've practiced versions of this speech a dozen times: on an acclaimed neurosurgeon with a renowned wine collection, an international judge who used to be president of the Spanish Wine Society, the director of wine research at the University of British Columbia and a dozen large commercial vintners—most of whom gave me a look like Michiel Kleiss is offering now: a mix of impatient skepticism, gentle dismissiveness and chagrin at discovering that they're dealing with a crackpot. But still, I think I'm on to something.

"But you . . ." interjects Kleiss's formidably cultured and charming girlfriend. "Do you have this problem? With migraines and wine?"

"Thank God, I do not."

"So, one less thing to worry about!" She lifts her glass. "Let's have another bottle."

JOINING THE THREE of us for dinner is Elard—the owner and operator of the Amsterdam Red Light District Tours. He is a smart young guy and somewhat in awe of our host, who is a consummate and, in fact, legendary one. Oddly, they haven't met until now, even though Kleiss's newest business is the main stop on Elard's Hangover Tour.

Michiel Kleiss's first business was a world-famous nightclub. The Roxy was to house music what Studio 54 was to disco or CBGB to punk. Party people still pilgrimage to where it once stood, while Kleiss has gone on to create two of the classiest, most coveted restaurants in the Red Light District. And now, in a city synonymous with both vice and progressiveness, Kleiss, the club-owner-turned-restaurateur, has become the healthy-yet-hedonistic, dashing, well-aged poster boy for a fascinating urban revitalization project.

Project 1012, named for the Red Light District's postal code, is a city-run initiative meant to make one of the world's most notorious, yet beloved travel destinations "safer, more attractive and more liveable" while not losing too much of the sex, drugs and rock 'n' roll. A pivotal part of the project involves closer regulation of the sex trade (purportedly to protect prostitutes) and buying up one-third of the storefront red-light window-brothels to turn them into something else.

"So, because I'm supposed to be the decent, upstanding entrepreneur of the 1012," says Kleiss, "they called me up and said, 'We want to give you a whorehouse to turn into something else. Do you have any ideas?' They couldn't think of anything themselves." And it *is* kind of tricky. Ninety percent of the businesses here are brothels, bars and THC cafés, so a bookstore or haberdashery might not do so well. But eventually, he did have an idea: "I think it actually came from the Project 1012's own literature," he says as the next bottle arrives. "Some slogan they had about 'a brighter future.' And that's when it hit me: hangovers!"

At first, the idea was just a friendly little shop that could provide things people associated with hangover relief: Aspirin, Alka-Seltzer and whatever other things Kleiss could think of. He took the idea to city council, and at the meeting, they loved it. But then, after a few days of perhaps all-too-sober second thoughts, they got cold feet—said it would be seen as promoting the kind of excess and vice the project was meant to curb. And that's when Kleiss lost his cool.

"I accused them of *trurrig*. Of, of . . . I don't know—being wet? A dingbell? No, that's not right. It is a great Dutch word . . ."

"It *is*," says Elard. "It is like soft and kind of stupid, but not exactly."

"Wimpy?" I say.

"Maybe wimpy!" says Kleiss. "But more . . . more . . . I don't know. It is such a good word, and it is not a good thing to be called."

I tell them I'll look it up.

"But the point is," says Kleiss, pouring out more wine, "they want to do this very tricky thing: to change this unique place, enough to make it cleaner, safer, to improve the economy, but not so much that it becomes something else, something more boring, and people stop coming. And they want to do that with what? Shoe stores? Cake shops?"

So Kleiss got pissed off and called them all something like wimpy but not quite. "Apparently that worked." he says. "But by then the idea had grown in my head—into something more interesting . . ."

And that, there, is when Kleiss started down the path that I've been on for quite a while now—in search of the elusive cure. "I looked and looked. There was a lot of stuff out there, but none of it made sense to me—and you've got to remember, I'm looking at it as someone who trained as a doctor."

"You're a doctor too?" I say to our club owner, restaurateur, entrepreneur.

"Well, no. But I *was* trained as one," he answers. rather confusingly. "The point is, the only thing in the whole world that I found and thought, *This might really work!* turned out to be a Dutch product. And not only that—it was also . . . how do you say it . . . up for grabbing!"

According to Kleiss, Reset had been created by a Dutch chemist with wealthy friends, spurred on by the prospect of week-long sailing regattas without worrying about all the drinking. And apparently, it worked so well that they invested in the product. But that didn't mean they could sell it.

"They went at it with the same boring ideas as everyone else," says Kleiss. "Women with big breasts, and get it free if you buy two bottles of rum, and target all the students. But it looked terrible, and tasted even worse, and nobody bought it. I mean *nobody*. And the investors wanted out." So there it was: the only thing in the world that an almost-doctor looking for a would-be cure thought might just work—and it happened to be up for the grabbing.

So Kleiss bought out the regatta investors and became, along with the chemist, a 50 percent owner of Reset. He redesigned the packaging, improved the taste from "not at all palatable" to "somewhat palatable" and was going to change the name to Brainwash, which he would also call the store. To a council committee trying to clean up local culture, however, the name was perhaps a bit too on point, or alien-overlordy. So they ended up sticking with Reset.

And as it turned out, they couldn't have named the store after the product anyway. In Holland, as in most EU countries, the laws around advertising a hangover remedy are extremely restrictive—insofar as you're not allowed to do it. What you can do, however, is "inform and educate."

"In the end, that was the best bit of luck," says Kleiss. "That we couldn't just open a store and market the stuff in the same old way. It's a complicated product, and we wouldn't have come up with this perfect way of explaining it to people if we hadn't been made to do so." And so, with the red light changed to blue, and windows full of Reset rather than half-naked women, the Hangover Information Center opened its doors to a brave new day.

WELCOME TO THE MONKEY HOUSE

I have looked it up: *swishy*, *sappy*, *fussy*, *frumpy*, *mimsy*, *dowdy*, *lame*, *boring*, *jerkish*. Put them all together, and you've got *trurrig*—a Dutch word that rattles cuttingly in a nation with its own official brand of bravery.

"Dutch courage!" says Gary King, hoisting a glass to face down the bluebloods in *The World's End*. "Like when the British soldiers used to drink Dutch gin before battles and it gave them super strength!" Which, of course, it did not.

Be it heroin, hash or hooch, the Dutch have long been in the business of creating ingestible courage for the rest of us. The Netherlands was the first European nation to develop large-scale distilling. And while foreigners claimed it was the shitty weather that made intoxicants such an integral part of lowland life, it was perhaps also something in their character.

"The innate Dutch inventiveness and love of experimentation," says Gintime.com, "meant there was virtually nothing they would not make into strong drink." The ships of the East India and Dutch West India Companies left Holland full of all kinds of booze and returned with the wealth of nations—and also lots of monkeys.

In 't Aepjen translates as "In the Monkeys," and it's the name of what is known as Amsterdam's oldest bar—or, at least, the place that's been serving jenever longer than anywhere else around. Jenever, made with grain alcohol and juniper, is the Dutch precursor to what we now call gin. If you're served it right—in a tulip-shaped glass that is ice-cold and filled to the brim—you're best not to pick it up. Instead, you drop your head and slurp it right off the table. When followed by a beer chaser, it's called a *kopstooje*, or "little head butt."

In the In 't Aepjen, they've been butting themselves in the head for half a millennium or so. It's still the closest inn to the old port, and where sailors back in the day bunked down for shore leave. Apparently, when they ran out of money, some of them paid their tabs with monkeys brought back from Southeast Asia—which is how the bar got its name. And also, people tended to get shanghaied here—that is, coerced or kidnapped into working on the ships. One moment, you'd be headbutting jenever, and the next you'd wake up in a room upstairs, hungover, half-devoured by fleas from all the monkeys, with a year-long seafaring contract pinned to your itchy chest.

And thus the Dutch saying *"in de app gelogeerd,"* or "to stay with the monkeys." It can be used for all sorts of trouble you might find yourself in, but most evocatively and precisely the kind where what looked like a good idea at the time came back and bit you in the ass like a monkey flea, butted you in the head like a dozen shots of cold, hard booze and then cast you to sea with a year-long hangover.

For example, it could be said that Steve Perkins had been "staying with the monkeys" when—after hours of drinking and late-night trading in which he somehow bought seven million barrels of crude oil and drove worldwide prices through the roof—he woke to a phone call from his employer asking where $540 million of the firm's money had gone.

The same could be said of Paul Hutton after he lost his driver's license for joyriding (albeit very slowly) his daughter's pink electric Barbie car while intoxicated at more than twice the legal limit. Said Hutton, a former aeronautical engineer, "You have to be a contortionist to get in, and then you can't get out. . . . I was a twit to say the least."

Then there's Alison Whelan, who, drunk on Lambrini, was heard shouting, "I'm Jack Sparrow" as she unmoored a forty-five-foot, double-decker ferry boat, smashed it into several luxury catamarans, then ran aground a mile upstream.

In 2002, Dutch researchers concluded that even slight intoxication reduces the brain's capacity to detect errors in judgment. But then, just recently, Dr. Bruce Bartholow at the University of Missouri challenged those findings by asking whether it's possible that the ability to detect mistakes or wrongdoing remains the same when you're drinking, but that when you're drunk, you just don't care as much. And then his research showed this to be true.

"It's not as though people do drunken things because they're not aware of their behavior," says Bartholow. "But rather they seem to be less bothered by the implications or consequences."

So, what if, in the moment, you truly want to be Jack Sparrow, an oil magnate or a Barbie race-car driver—or just pay your tab with a monkey—and by getting more inebriated, it all feels much more possible, tomorrow be damned. Well, then, perhaps the trick is to simply embrace your drunken decisions—like Sam Smith did.

Mr. Smith, who happened to have the same name as a somewhat annoying pop star, got drunk with his friends one night and agreed to change it—to Bacon Double Cheeseburger. He filled out the appropriate application forms, and then, when the paperwork arrived a few weeks later, made sure he was drunk again to complete the process. "I have no regrets," said Cheeseburger to reporters. "My mum was furious, but my dad thinks it's hilarious. He's more than happy to use my new name."

Whether he knows it or not, Bacon Double Cheeseburger is following the ancient Persians in their traditions of thinking and drinking. According to Herodotus: "If an important decision is to be made, they discuss the question when they are drunk . . . Conversely, any decision they make when they are sober is reconsidered afterward when they are drunk." Similarly, the historian Tacitus wrote of the ancient Gauls that only after getting sufficiently drunk would they "deliberate on the reconcilement of enemies, on family alliances, on the appointment of chiefs, and finally on peace and war; conceiving that at no time is the soul more opened to sincerity, or warmed to heroism."

In fact, such thought persists throughout modern philosophy. Immanuel Kant insisted that the "openheartedness" created by drinking was "a moral quality." But he also warned that "stultifying drunkenness has something shameful about it." Which brings us back to monkeys, and a Brazilian mechanic whose openheartedness extended, perhaps, a bit too far into the animal kingdom.

João Leite Dos Santos, from São Paulo, Brazil, was captured on camera swimming across a pool at the Sorocaba Zoo to drunkenly befriend a group of spider monkeys. In the clip, the little primates can be seen swarming Dos Santos, who eventually had to be rescued by onlookers, then taken to the hospital with severe bite wounds—providing us, finally, with a modern-day, non-metaphorical example of staying with the monkeys—and also trying to get one off your back.

THE FUTURE'S SO BRIGHT (YOU GOTTA THROW SHADE)

The dark alleyways of the Red Light District are so narrow, your breath mixes with that of strangers as you pass; so dark, you'll never know their faces. It is an ancient sort of intimacy—feral, fetid and full of mystery. There are drugs in the doorways, sex in the windows, a thousand years' worth of hopes and dreams dropped between the cobblestones over which you stagger, toward the cold blue beacon at the end of the street. The sign on the door reads: HANGOVER INFORMATION CENTER: "FOR A BRIGHTER TOMORROW." And stepping inside really is like moving from a shadowy past into a crystalline, Kubrick-esque future: a long, bright, white room made longer in the mind by the trick of mirrors—a stretching white counter and a thousand (maybe ten thousand) glowing blue bottles in endless rows on the wall behind. The man at the counter fits the room impeccably: long and tall, in a crisp blue shirt, with silvery hair and steel-blue eyes. "How do you feel?" says Michiel Kleiss.

This isn't my first time in Amsterdam. There is an ancient, layered debauchery here that grows more profound with every little spark it swallows from lustful souls across the globe. As opposed to a Vegas hangover—which hurts, for sure, but in a sort of antiseptic, modern way—an Amsterdam hangover wrestles with your bones and licks at your lizard brain like it was painted sideways by Hieronymus Bosch.

And considering the past couple nights—all the drinks and

other things, two dozen bars, an impromptu bartending gig, the Red-Light Hangover Tour, forgetting where I was staying and the stint in the monkey house—my brain feels almost okay, or at least the right way round. And I'm willing to concede that this may be due—at least in part—to the Reset. But as Kleiss pointed out, it is a complicated product.

Even just taking Reset requires concentration and a little bit of dexterity: each 400-milliliter cylindrical plastic bottle of blue liquid is capped with an upside-down 100-milliliter bottle of white powder. At the appropriate time (as close as possible to that precarious breach between drinking and sleeping), you cut through the plastic wrapping, separate the two bottles, open both, pour the white powder into the blue liquid, replace the cap and shake it. So far, I have taken it once in Kleiss's restaurant at the end of that boozy dinner; once between puffs of something or other, with my feet dangling over a canal; and once among the monkeys. When dissolved, it tastes both bitter and sweet and has the attributes of a sycophant: cloying, dubious and surprisingly helpful.

The key ingredient—the one that apparently made our man Kleiss sit up and take notice, and also the one that requires such gymnastics regarding delivery system and flavor adjustment—is glutathione: an antioxidant found in fungi, plants, animals and people. According to the Hangover Information Center, glutathione mitigates a hangover by lowering levels of acetaldehyde. And as opposed to various other groundbreaking remedies, the literature insists that taking Reset will not mess with your buzz. "You will remain as drunk as you are," reads an information poster on the wall. "But tomorrow will be a lot brighter."

"So you survived okay," said Kleiss. "And now you go back to Canada?"

"Soon," I tell him. "First, I'm having coffee with Dr. Verster in a town I can't pronounce."

"Oh, right," says Kleiss. "The *real* doctor." He gives me a knowing sort of look, and a bottle of Reset for the road.

AS OPTIMISTIC a vibe as one gets from raising a glass with Michael Kleiss, drinking a cup of coffee with Dr. Joris Verster is a sobering affair. In some ways, these two men can be seen as the two extremes of hangover research in Holland.

Kleiss, after all, is a slick and charming entrepreneur who makes no bones about his business interests. He is approaching the morning after from a top-down perspective. Having found a product that interested him, he set about acquiring and improving it and is now attempting to demonstrate and market its efficacy. In so doing, he has created the Hangover Information Center.

Verster, meanwhile, is a scientific academic who has authored more hangover studies than even I am willing to read. He is widely recognized as the world's leading expert on hangover research, as well as the creator and de facto head of the Alcohol Hangover Research Group. And yet, he appears to have no interest in Kleiss or Reset whatsoever.

In fact, without either of them mentioning it, I have discovered that Kleiss has been courting Verster with letters and samples of his product—to which the doctor presumably responded in much the same way as he does when I finally meet with him and promptly bring it up: "There is no scientific evidence to suggest that any of that is effective at all. There have been no studies."

But wouldn't the Alcohol Hangover Research Group be the perfect entity to conduct such studies? Verster doesn't see it that way. His is the bottom-up approach—studying hangovers from the perspective of an underlying cause. And until *that* research is complete, he sees no purpose at all in looking at something like Reset.

Right now, Verster is of the mind that much of what we experience as hangover is due to both an autoimmune response and inflammation. This jibes with the thoughts of other smart people I've been talking to, including Dr. Spector of the British Gut Project, *Proof* author Adam Rogers and more than a couple of product marketers.

But when I suggest to Dr. Verster that studying certain products, substances, chemicals or ingredients that appear to be effective in mitigating aspects of hangover might, in fact, lead to hints about an underlying cause, not to mention the negation of very specific symptoms, his dismissiveness is so complete that I feel a late-day hangover blooming on the edge of my brain.

ON MY WAY back across the ocean, poring over papers, pamphlets and my own notes, I finally come to the realization that Kleiss's glutathione and my N-acetylcysteine (NAC) are pretty much interchangeable. Or rather, NAC is a precursor to glutathione. Without the somewhat awkward equation of *powder + liquid = bad-tasting elixir*, there is no easy way to get enough glutathione into you directly—but higher doses of NAC in capsule form will then manufacture lots of glutathione. At least, I *think* that's how it works.

I have been dialing in on this curious amino acid more and more. But looking through websites and the packaging of various new hangover products that use NAC, the explanations as to how it or glutathione might help with a hangover are varied and confusing. Put most simply, if it works, it would do so in a way that minimizes, if not stops, the chain reaction of my system in a state of alcohol withdrawal by protecting my body's cells, breaking down acetaldehyde and mopping up free radicals—which could be seen as dealing with the beginning, middle and end of the beginning of a hangover. Or something like that.

Perhaps most surprising, at least to me, is the fact that egg yolks contain NAC. They may not be as potent as the capsules I've got, but it does lend credence to all those ancient owl-egg recipes attributed to Pliny. And also the validity of a classic British breakfast.

Of course, as Verster is quick to point out, nobody has done studies to prove such things. In fact, in looking at what's out there, I am beginning to think my own tests—on family, friends and acquaintances, but mostly on myself—are among the most extensive.

One new product that's caught my attention is a mix of all the most recent heavy hitters in the realm of hangover supplements. Party Armor, manufactured in Detroit, appears to combine prickly pear, Japanese raisin and NAC with several other ingredients. So I think I'll be heading to Michigan soon.

And at the same time, thanks to Kleiss, I have plans on how to tweak my own NAC concoctions—with higher doses ingested after drinking and before sleeping, rather than before drinking and then again after sleeping, as most products containing NAC suggest. In fact, I'm becoming ever more sure:

By the next day, it will always be too late.

KILLER PARTIES

Today, there is no debate; historians, humanists, economists, evangelists, poets, politicians and pretty much everyone agree that Prohibition in America was an awful idea executed terribly, achieving the opposite of its intentions. Crime intensified and became organized, daily life became more dangerous, jails filled up, and the economy suffered. There was more corruption, more divisiveness, less faith in government, less respect for authority, more lawlessness, more dangerous drugs on the streets, more alcoholism and a decade of the worst hangovers ever—not to mention psychopathic Santa Claus hallucinations.

On Christmas Eve, 1926, a man burst into New York City's Bellevue Hospital, shouting that Santa was coming to kill them all with a baseball bat. Before the nurses could figure out what the hell was going on, the man dropped dead. Then another yuletide yowler came stumbling in, and another, and another. By the time Christmas was over, hundreds of New York revelers were going though hell in the hospital, and more than thirty had died.

Although this event was bizarre and horrific, medical practitioners were already somewhat accustomed to this kind of thing, due to what economist Richard Cowan calls the Iron Law of Prohibition: "When drugs or alcoholic beverages are prohibited, they will become more potent, will have greater variability in potency, will be adulterated with unknown or dangerous substances, and will not be produced and consumed under normal market constraints."

At the beginning of Prohibition in the United States, the majority of illicit booze was good old Canadian whisky, smuggled across the Detroit–Windsor border. We Canadians had our own, uniquely profitable, form of Prohibition: though liquor was illegal to buy or drink, we could still make and sell the stuff to Americans—at rum-running prices. By the mid-'20s, however, crackdowns on cross-border smuggling had pushed other, more local, methods of production into high gear. But makers of American moonshine couldn't risk properly aging their product. So, as Iain Gately puts it, "They added dead rats and rotten meat to achieve the same effect."

And then there was the hooch made from industrial alcohol, which the Volstead Act had required to be "denatured"—the government word for adding crappy-tasting chemicals to render it undrinkable, or at least extremely unpalatable. But as Deborah Blum, author of *The Poisoner's Handbook*, points out, "The bootleggers paid their chemists a lot more than the government did, and they excelled at their job."

In fact, they "renatured" so much swill that, by 1926, stolen-then-redistilled industrial alcohol had become the primary source of booze in the US. In a sort of chemical-weapons race, industrial manufacturers were forced by the government to up the toxic payload even further. And so, with a brutal logic that could have seemed sane only amid the madness of American Prohibition, the US government began effectively, and knowingly, poisoning its own citizens.

"By mid-1927," writes Blum, "the new denaturing formulas included some notable poisons—kerosene and brucine (a plant alkaloid closely related to strychnine), gasoline, benzene, cadmium, iodine, zinc, mercury salts, nicotine, ether, formaldehyde, chloroform, camphor, carbolic acid, quinine, and acetone. The Treasury Department also demanded more methyl alcohol be added—up to ten percent of total product. It was that last that proved most deadly."

The Christmas Eve killings were just the beginning. By the end of Prohibition, "the federal poisoning program," as Elum refers to it, had murdered as many as ten thousand people. Although practically forgotten today, it was public knowledge at the time. In a 1926 press conference, New York City medical examiner Charles Norris warned people of the threat and was unequivocal about the blame:

> The government knows it is not stopping drinking by putting poison in alcohol, yet it continues its poisoning processes, heedless of the fact that people determined to drink are daily absorbing that poison. Knowing this to be true, the United States government must be charged with the moral responsibility for the deaths that poisoned liquor causes, although it cannot be held legally responsible.

Or, as the fast-talking, rope-tricking, movie-star cowboy and columnist Will Rogers put it, "Governments used to murder by the bullet only. Now it's by the quart."

Prohibition corroded America's view of itself—at least for those white and well off enough to still be surprised by the underbelly. When gentleman bootlegger George Cassiday—who for ten years supplied top-shelf booze to the very congressmen who had voted for Prohibition and helped poison their fellow Americans—spilled the beans (but without names) in a series of articles for the *Washington Post*, the hypocrisy and abuse of power was too black and white to ignore. As medical examiner Norris pointed out, the people dying were those "who cannot afford expensive protection and deal in low-grade stuff."

But rival gangs and Republicans weren't the only ones holding killer parties. Thanks to true American moxie, and the divine likes of Dionysus and Dizzy Gillespie, the Prohibition era was also the Jazz Age, the war on booze was the resurrection of Bacchus, and Americans partied like never before.

In every major city, there were soon twice as many speakeasies as there had ever been saloons, and they were a helluva lot more fun—in an all-inclusive, partners-in-crime kind of way. When just having a drink could put you in jail, or even the morgue, then you'd better damn well enjoy it. Cocktails became the drink of choice as barkeeps found new and inventive ways to mask the taste of dubious spirits. And now men and women, black and white, drank together, dancing to music they'd never known existed. The stakes were higher, and so was the populace. The growing black market and a new taste for the illicit had made street drugs like cocaine, marijuana and heroin suddenly much easier to come by.

And so, with new drinks, new drugs, new music, men and women of all creeds, colors and classes smoking and drinking and dancing all night in acts of hedonistic protest and rebellion, the modern booze-up was born—its hangover then brought to bear on a grand scale in the Great Depression. But this renaissance in partying—a sort of democratic debauchery in the face of The Man—carried forward ferociously, from the flappers to the beatniks, the Blues Brothers to the Beastie Boys, to the all-night party kids and on into the morning.

PART EIGHT

THE TIGER ON THE ROOF

IN WHICH OUR MAN ON THE BORDER CROSSES A RIVER,
DRINKS IN DETROIT, THEN WATCHES THE SUN RISE WITH
A *PLAYBOY* MODEL AND A PISSED-OFF BENGAL TIGER.
APPEARANCES BY BASEBALL'S GREATEST LEGENDS,
AMERICA'S RICHEST INDIAN AND ONE VERY BIG CAT.

Beneath every illness is a prohibition.
A prohibition that comes from a superstition.
—ALEJANDRO JODOROWSKY

CROSSING FROM WINDSOR, ONTARIO, CANADA, into Detroit, Michigan, USA, is usually a bit of a head trip, with ominous if not apocalyptic overtones. And today is no exception.

As I approach the border, the sky ahead is filling with a column, then a cloud, of thick black smoke, the kind that never bodes well in any situation. And then I come upon it: an inexplicably exploded, flame-engulfed SUV on the side of the highway; the cars just roll by, not a bazooka-wielding super-villain or a fire truck in sight. I pull over, walk up close enough to ascertain that mine is the only human body anywhere around, then get back in my rental and drive into America.

This stretch of river I'm crossing over has seen plenty of weird and dodgy stuff. During Prohibition, it is estimated that over half the massive amount of illegal booze bought, sold and drunk in the US came across, through or under this very river: over the ice on single dogsleds, in midnight convoys of delivery trucks, hoisted along cables through the rooms of sunken houseboats, and even pumped through a few short-lived pipelines running all the way from Canadian distilleries.

To make matters even more colorful, Detroit rum-running and the associated crime syndicate was controlled by the Purple Gang:

a group of charismatic Jewish badasses whom even Al Capone wasn't willing to mess with, so he partnered with them instead.

Part of that was due to Detroit's strange position as harbinger of doom. Recessions, depressions and even restrictions tend to hit here first and fiercest. Prohibition came to Michigan in 1917—thanks mostly to Henry Ford, who wanted a sober workforce in his Detroit plant—so that when the Volstead Act went nation-wide, gangsters here were ahead of the game and rum-running had already become Michigan's second-largest industry after automobiles.

The other part was the Purple Gang itself. It is not clear how they earned that name, but the longer they ruled the city, the more horrifically violent and obsessively flamboyant they became—the prototypes of a uniquely Detroit kind of gangster: a strong, soot-soaked songbird, singing and spitting beneath the streets. And of course, no matter how serious things got, it was also always for sport. This is the city, after all, not only of boom and bust, booze and prohibition, but of Lions and Tigers and Red Wings too.

Per capita, Detroit saw more brutality and bandit booze during Prohibition than anywhere else. At the height of US rum-running, New York City, with a population of five and a half million, had thirty thousand speakeasies, gin joints, booze cans and bolt-holes; meanwhile, in Detroit, with fewer than a million people, there were twenty-five thousand. And here the hooch was still pretty good—as opposed to the thinned-out, cut, then chemically filled-out products that would eventually make it to the heartland and beyond, or the moonshine with more methanol than ethanol, or the refurbished rat juice made from industrial alcohol.

So, while the '20s were the worst decade for hangovers in the history of America, not so for the hard-drinking denizens of Detroit, who were pounding back the good stuff. It is an irony even more acute since the city itself has become the embodiment of metaphorical hangover. It was just 2013, after all, when Detroit declared bankruptcy, houses downtown were being sold for a dollar and newscasters quoted *Robocop* on the evening news.

Economically, structurally, socially and historically, the Motor City is seen as a warning of what's on the other side the demise, the must-come-down, the morning after. But now, Detroit is experiencing the morning after the morning after—and, as such, it's the perfect place for the business of hangovers, at least theoretically.

DRIVING INTO DOWNTOWN is tricky these days—not because of mayhem and robots with guns, but rather an abundance of construction and not enough signage. It is being rebuilt quickly but purposefully. As with Amsterdam, entrepreneurs, developers and businessmen are being given all sorts of creative incentives to help revitalize the downtown core. And in fact, I'm here for much the same reason that I was there: to test out a brand-new hangover product over dinner and too many drinks with the guy who owns the company. But that's where the similarities end.

Whereas the Dutch Reset bottle is somewhat futuristic—in a crystalline kind of way—Party Armor Protection comes in a gunmetal black cylinder, made to look like a shotgun shell: "The last shot of the night." And whereas the tours I took in Amsterdam focused on bordellos, bars and modern art, here it's all gangs, cars, Prohibition and professional sports.

THE MIDDLE BALL

That Al Capone and Babe Ruth remain the two most enduring figures of 1920s America speaks to both the legacy of Prohibition and the wonderment of baseball. It is, after all, one of sports' profoundest anomalies that the most prolifically drunken gluttonous, overweight and seemingly unhealthy player to ever grace the field was also one of its greatest. "It's simple, kids," counseled the Babe. "If you drink and smoke and eat and screw as much as me? Well, kiddos, someday you'll be just as good at sports."

Babe Ruth was so fundamentally robust that it seems not even hangovers could get to him. When the Yankees came to Chicago on a road trip, the White Sox enlisted their favorite bartender to pour as much of his infamously powerful punch as the Bambino could possibly drink, which was a lot. As his teammate Tommy Henrich put it, "that guy has a throat like a trombone." Ruth got fully smashed and then some, stayed up till dawn, trounced the White Sox, then crossed the field to ask where to meet them for drinks that night.

But it's not like the Babe was the only slugger to hit the bottle. Many others have played their best when at their very worst. Though he'd once graced a box of Wheaties, the great Mickey Mantle admitted in *Sports Illustrated* that his actual "Breakfast of Champions" was brandy, Kahlua and cream. Once, just three weeks after breaking his ankle and halfway into a road trip (thus allowing for a full-on binge), he was crashed out on the bench in Baltimore, nursing another hangover, as well as that stupid ankle—or so he thought. In fact, the team doctor had cleared him that morning—and in the bottom of the seventh, he was called on to pinch-hit.

According to biographer Allen Barra, "Mantle later said he saw three balls coming at him and decided to swing at the one in the middle." A home run, of course.

Just as that quote sounds too good to be true, it's also too good to be used only once. As Bert Randolph Sugar writes of Paul Waner, a slugger from the generation between Ruth and Mantle: "If the *Macmillan Baseball Encyclopedia* ever listed 'hangovers' among its many statistics, then the all time leader in that category would easily be Paul Waner. For no one ever nursed more than Waner, who once, when asked how, if he was so hung over so often, he ever managed to meet the ball, answered, 'I see three and hit the one in the middle.'"

Waner, with the awesomely toxic nickname Big Poison, also had a uniquely ambitious yet straightforward hangover remedy: lots and lots of backflips. According to his roommate with the Boston Braves, Buddy Hassett, he'd do "fifteen or twenty minutes of backflips and he was cold sober, ready to go out to the ballpark and get his three hits."

Then, of course, there is David "Boomer" Wells—a modern-era attempted reincarnation of the Great Bambino himself. Blessed—or cursed—with such similar heart, liver, lust for life and general physique, he was fined $100,000 for sporting Ruth's old Yankee cap on the mound, and he still wears his devotion on his sleeve, or rather, just beneath it: an intricate rendering of him pitching to the Babe tattooed across his throwing arm.

After stints with the Toronto Blue Jays, Detroit Tigers, Cincinnati Reds and Baltimore Orioles, Wells ended up in New York, where he asked that Ruth's number 3 be unretired so he could wear that too. Instead, they gave him 33.

Finally a Yankee, the hard-drinking, fast-living, somewhat quirky Wells Ruth-ified the rest of his persona—hitting the town night after night with a gleeful ferocity that would have made his hero proud—or at least amused. By the spring of 1998, Wells was famous, or infamous enough, to be hosting *Saturday Night Live*. He drank at the cast party until the sun came up, got one hour's sleep and stumbled onto the pitcher's mound for an afternoon game against the Twins.

No matter how difficult it is to hit a good pitch—never mind with a hangover—it is practically impossible to throw one, over and over and over, without anyone ever hitting it, or even getting on base, for a whole nine innings, a feat called a perfect game. Until that day—May 17, 1998—only fourteen players in the entire history of Major League Baseball, dating all the way back to 1880, had ever accomplished this.

And then, beyond all possibility, let alone comprehension, David "Boomer" Wells, in a stadium full of screaming children (thanks to a well-timed Beanie Baby promotion) pitched the fifteenth-ever perfect game while, in his words, "half-drunk, with bloodshot eyes, monster breath, and a raging, skull-rattling hangover."

THE TIGER ON THE ROOF
(IN THE FOREST OF THE NIGHT)

I've been to some weird dinner parties in my time, and this one's right up there. Around a large table in a private dining room in downtown Detroit's most opulent restaurant, I am sitting across from a Russian heiress who breeds racehorses in Upstate New York and poses for *Playboy* for the hell of it. Next to her is a tattooed, Rasputinesque guy who just parked a convoy of cages containing a bobcat, two wolves and a Bengal tiger that he drove all the way from Colorado. Then there's the wild animal trainer, the wranglers, a couple more models, our host, his entourage and Cason Thorsby of Party Armor—who seems nearly as baffled as I am.

We are here as guests of David Yarrow—an internationally famous English photographer whom Thorsby happened to meet at a club last night. We are also invited to his photo shoot at dawn, where he will attempt to somehow capture the people at this table, the wild animals parked on the street and the sun rising over a burned-out, abandoned Detroit factory.

"So, what do you think?" asks Thorsby, opening his hand as if to catch the room. He'd been trying to come up with a unique way for me to test his Party Armor. And then it was just sort of tossed to us: this surreal cinematic mash-up of an Andy Warhol "happening" and the *Hangover* movies, complete with a Russian heiress, a rooftop film crew and a tiger on a leash.

"Oh, this'll do," I say as we finish our bourbon sours and start on a bottle of red.

Yarrow stands to address the table. He is a broad-shouldered man with a steady gaze and commanding voice. He welcomes us, talks a bit about the shoot and ends by saying: "Please enjoy your meal, but don't have too much fun tonight. I need you all at your best when the sun comes up." Thorsby and I clink our glasses together. What could possibly go wrong?

CASON AND HIS younger brother Sheldon came up with Party Armor as they were finishing college and Detroit was going bankrupt. "We drank like you do in university," says Cason. "Most guys were fine, but for us, the next day was just brutal. And also we wanted to stay here and stick by our city. So it just seemed like the right thing."

Maybe not the clearest correlation, but Cason has always been entrepreneurial, and his brother—a chemistry major who was into health and fitness—knew a lot about dietary supplements. So while Cason cased the marketplace, Sheldon searched for something to sell. He found there was a lot of stuff out there that looked good for treating hangovers and wondered what would happen if you put them all together. For his part, Cason looked at consumer reports and market research, then called up the modern guru of supplements.

"Manoj Bhargava. He lives right here in Michigan," says Cason, with a note of reverence, as if I should know who he's talking about. "Five-Hour Energy? The richest Indian in America?"

As it turns out, that is how Bhargava refers to himself—at least in a recent *Forbes* cover story. Those little bottles of Five-Hour

Energy are a ubiquitous, massive worldwide seller, and yet few people realize they began as a sort of chaser to a hangover remedy called Chaser.

"Chaser was a big deal. Not like Five-Hour Energy became, but it got him started."

So Cason figured Bhargava was just the guy to talk to about launching his own hangover remedy: someone local, who'd already done it and done it well. "I called him up, out of the blue, and just started asking all the questions. And the two things he told me about hangover remedies have already proved true: everyone is skeptical, and everyone is an expert." Cason puts down his wine and picks up a glass of water. "They all think it's about dehydration, which would be so easy to solve. And at the same time, people think it's unsolvable—that if there was actually something to the product, they would already have heard of it. It would be on every front page and I'd be on *20/20* as the most interesting man of the year—otherwise, it's got to be a scam."

For Cason, it's more about the psychology of the consumer and the availability of the product than the product itself: "It has to be as available as booze. That's the real thing that Bhargava succeeded in. He got Chaser into stores everywhere—Walmart, Walgreens, you name it—and it became the most successful hangover product ever. But even that just made him $12 million, at least according to reports, and they could be wrong; it turns out they left a lot of product on shelves and had to buy it back. So who knows how much he really made, or even lost? But it's moot anyway. The guy's a billionaire now. He became the richest Indian in America by taking hangovers out of it all, and just selling five hours of energy—whatever that even means. It's fucking impressive, man!"

Undeterred by the guru's warnings about consumer skepticism and laziness, the Thorsby brothers charged into the spot left by Chaser with their Party Armor on. Sheldon had found a way to put a whole whack of stuff, including NAC, glutathione, prickly pear, milk thistle and all sorts of vitamins and minerals into a 2¼-fluid-ounce bottle—and it didn't even taste too bad.

That was another thing revealed through Bhargava's billion-dollar supplement jiujitsu: consumers shied away from pills, but they didn't mind taking a shot. And Cason, for one, was going to take his—learning the lessons of others to make his own go of it.

AS DINNER WINDS down, the others go back to their cages and rooms while Cason and I go out on the town. We meet up with a few friends of his in a bar. They're all affable guys who swear that Party Armor Protection works for them every time. The Thorsby brothers also make Party Armor Recovery—in a white bottle, to be taken the morning after. The two products stand, like white and black shotgun shells, on the table among our tumblers.

"They both work, and work together," says Thorsby. "But if you're going to just take one, make it the Protection, and before you go to bed."

"In my experience," I tell him, "by the morning it's too late."

"Yeah," he says, and picks up the white bottle. "But this is still a pretty cool product."

I have questions for Thorsby about the ingredients in both—or rather, their quantities, strengths and percentages—but he's not able to tell me much. It's not that he's trying to protect the formula; he just really doesn't know. His brother, after all, is the chemist in the equation, and he is nowhere to be seen.

After a few more drinks, Cason reveals that he and Sheldon have "had a falling out" over Party Armor. He won't get into the specifics, but is obviously bothered by it.

We are supposed to be up on a roof with a tiger before the

sun rises. So Cason suggests we do our "last shot of the night" at my hotel bar, where, with another double Jameson, we open our little black bottles and swig them down. It has a lemony-sweet and acrid taste; not so bad at all. Then Cason heads home to sleep, planning to pick me up in just a few hours.

I should crash out too. But I also know that, thanks to recent Senate approval, last call in Detroit isn't for another three hours, and there are probably some historically interesting bars in this neighborhood. At this point, I'm going to be waking up still drunk rather than hungover anyway, so I might as well see how Party Armor protects against that. Cason's given me a whole case of the stuff to bring back and try out, so there's plenty of time for less dramatic tests in the future. And also, I could just really do with another drink. It's been that way lately.

IN THE PAST two months, since returning from Amsterdam, I have tested more products than I did in the previous year. At home I now have half-empty cases of Pretox, Drinkwell, Hangover Gone and Sobur—all compliments of their makers. Some have worked fairly well; some not at all.

At the same time, I've been tweaking my own concoction to the point where it now appears to be outperforming everything else I've tried. I've suspected for a while that N-acetylcysteine (NAC) is pretty much my magic ingredient, and my conversations with Kleiss helped me realize the next step: upping the dosage. And, as far as timing goes, taking a cue from the instructions of products like Reset and Party Armor, making sure to take it in that precarious zone between drinking and sleep.

I am also pretty much drunk every night now—which may be the most telling, predictable and somewhat scary indication that I'm truly on to something.

🍾 🍾 🍾

THE TASTE OF FREEDOM

Ironically, Prohibition in America ended up making alcohol more socially acceptable and accessible to those from whom, historically, it had been most prohibited.

Under slavery, African Americans risked corporal punishment, even death, if caught drinking. Such has been the way in all societies that owned slaves—with the notable exception of the Spartans, who often got slaves obscenely drunk to use their sickness in hangover as a cautionary spectacle for the masses.

And of course, through much of human history, women have also been viewed as property. In the American land of the free, it took all kinds of civil wars, wars abroad and Prohibition to even begin to recognize a woman's God-given right to get drunk. So it's fair to surmise that, as bad as it's been for anyone to be hungover, for women in general, it's usually been worse.

Some scientific studies suggest that women are more prone to hangovers than men, while others suggest otherwise. But really, they're just looking at the physical hangover. When it comes to the metaphysical one, historically, there'd be no comparison—since, more often than not, women have been literally shamed for drinking, let alone getting drunk. So, probably best to hide your hangover, at least among the pious misogynists.

Family therapists Claudia Bepko and Jo-Ann Krestan write that for men, "out-of-control behavior associated with drinking historically has been accepted, even encouraged," while women who drank were seen as "sick, evil corrupters of men and children." And of course, it's always had to do with control and sex: "In first-century Rome, drinking by women was a crime. A woman suspected of drinking was assumed to have betrayed her husband sexually and could be put to death."

Such inscrutable logic (if not always such a harsh penalty) has been passed down through the ages in almost every culture. As a thirteenth-century poem by Robert De Blois surmised,

She who gluts more than her fill
Of food and wine soon finds a taste
for bold excess below the waist.

Conversely, in "Water with the Wine," from her 1976 self-titled masterpiece of an album, Joan Armatrading confronts the trauma of date rape at the hands of a drunken man. In the painful morning light, the dawning of memory becomes a waking nightmare as she turns her head to find her attacker still there, snoring on a pillow.

The writer and singer of the song—no matter how brilliant and resilient—has little recourse but to make a mental note, harkening back to Plato, to next time mix the wine with some water. The resigned tone in which Joan Armatrading delivers this sort of self-admonishment is indicative of how terrifyingly normalized trauma can become in a patriarchal society.

Even outside the fraught realm of sexual violence, it is difficult to find precise historical accounts or depictions of women's hangovers. As my dear friend (and a great writer) Tabatha Southey wrote to me, "Women have left fewer words everywhere, about everything, and aren't as likely to boast about their hangovers as are men. A man's hangover can be a kind of souvenir from his exploits in a land women have historically been unwelcome to explore. We skulk in, we skulk out. It's not funny when we're drunk or hungover; there was never a female Falstaff."

As recently as 1987, the brilliant French novelist and film-maker Marguerite Duras wrote of how her own drunkenness was perceived: "When a women drinks, it's as if an animal were drinking, or a child. Alcoholism is scandalous in a woman, and a female alcoholic is rare, a serious matter. It's a slur on the divine in our nature."

As such, it is not surprising that self-portraits of the hungover female artist are hard to find in any form. But there are a few famous ones by their male counterparts—often lovingly drawn. As a crippled, alcoholic dwarf who spent most of his time in Parisian bars and dance halls, Henri de Toulouse-Lautrec might have used his own visage for a painting entitled *The Hangover*. But instead, he put Suzanne Valadon's morning after onto the canvas.

Valadon, his drinking companion and fellow artist, had been a talented circus performer until she tumbled from the trapeze at the age of fifteen. And though neither of them was much older than twenty when Toulouse-Lautrec captured this particular morning, you can see in her eyes the sedate knowledge that her high-flying hopes are a thing of the past, and whatever stretches out ahead is something that's hard to hope for.

THE TIGER ON THE ROOF (THY FEARFUL SYMMETRY)

As the sun rises eight floors up, onto the rubble-strewn roof of this burned-out factory, the shapes are revealed in the glowing half-light: a tall blond woman, naked beneath a black-belted leather coat, holds in her hands a thick chain, at the end of which a Bengal tiger, its striped coat seemingly aflame in the dawning light, stands giant and huffing at her side. A dozen men in ball caps, most of them dark-skinned, are perched in various positions among the graffiti-covered smokestacks and piles of concrete and rebar, looking down at the woman and the tiger, who are stalking the sun across this mangled rooftop.

I do not know what this is supposed to be about: race and gender? Control and powerlessness? The chains that bind? Sex and stereotypes? Or maybe nothing at all; just a hazy hungover fever-dream.

The higher the sun rises, the more volatile this space becomes, and things seem to be falling out of control, out of frame; the tiger growing restless, then angry, ready to stage some fundamental protest; and now the whir of helicopters—police and news crews circle overhead . . .

HOURS LATER, BACK in Toronto, the story—if you can call it that—is all over the evening news: "Tiger gets loose during photo

shoot in the ruins of Detroit factory." There is footage—some
from the air, some from the ground, and then in a crumbling
stairwell—of a tiger who's sick of being told what to do. A hand-
held camera documents his scared stubbornness, as a tattooed
Rasputin shakes lawn tools in his direction.

It is hard to know what is what: sunrise or a dream; future or
fiction; the search for a cure or the thirst for a cause. I don't even
know if the Party Armor worked. Drunkenness, sobriety and
hangover have all begun to blur. But at least for now, I am home.
The world is still with us. And the tiger is back in its cage. I put
down my empty armor and pick up a bottle of something else.

I WOKE UP THIS MORNING

With an awful aching head. My head in the dirt. A buzz rollin' round my brain. My heart on fire. With a burning fever. The sun shining in my eyes. Bitterness in my mouth. A trail of teeth under the door. A wineglass in my hand. Flies in my beard. Alone on the floor. Feeling quite weird. At 11:11. To an empty bed. To an empty sky. In my clothes again. Trouble knocking at the door. Butter and eggs in my bed. My dog was dead. Blues around my head. The blues sitting on my face. And my baby was gone.

That last one is the quintessential one, of course—so fundamental and simple in its dawning heartbreak, like it's always been there, and always will—the original blues:

I woke up this morning, my baby was gone
I woke up this morning, my baby was gone
I'm heartbroke, hungover, and I'm all alone.

Those twelve-bars from the Mississippi Delta, the cotton fields back home, were known, quite simply, as the "I woke up this morning" blues. And as they stumbled raggedly north, through the Great Depression, the dead dogs, lost loves and repossessed ranches of bluegrass and honky tonk, all the way to Chicago and Detroit, they became a raging American hangover—now known as rock 'n' roll.

Sure, sometimes in song, people wake up feeling just fine, madly in love, and it's a sunshiny day. But far more often, not at all. Bessie Smith woke up alone. Robert Johnson couldn't find his shoes. Warren Zevon fell out of bed. Mark Knopfler's Jacuzzi was broken. And King Missile had misplaced his penis—so he had to get up and figure things out.

Part of the idea of "I woke up this morning," after all, is to then proceed with the rest of your day, often starting with breakfast—or lack thereof, as in the brilliant Chicago song "An Hour in the Shower—A Hard Risin' Morning Without Breakfast."

For those who only know the later Chicago—radio-friendly, movie-soundtrack Peter Cetera hits—this isn't that at all. It is growling, whimsical and unpredictable—penned and sung by the late, great, heavy-drinking Terry Kath, who lived life hard and ended it haphazardly, with an accidental bullet to the head. His songs were haplessly willful and willfully hapless. And of course, to read the lyrics isn't enough; you've got to listen to the song—mostly the beginning, before his day spins out into jazzy, spinning, prog-rock chaos.

But read it too and see what Kath has done: syntactically reversing the traditional, seminal line, calling out the morning blues, the hangover, and looking it straight in the eye. It is like a drunken trickster writing brave and slippery poetry, with the voice and chops to pull it off:

Today when I awoke
The morning blues hung over me
So . . . I looked it straight in the eye
I jumped into the shower for about an hour

Aw it was fine
Yeah, it helps me all the time
It's soothin' to my mind
Just to see those blues
Go slippin' down the drain . . .
Now, I usually have my breakfast
Which consists of tasty Spam
Yeah, I could eat it all day long
But I only love one brand
And I can't find it way out here
So . . . I have to take a pass
And settle for some hash
What a drag you're not here
Oh . . . sweet, sweet Spam.

This hankering for Spam, achingly rendered, is also part of the "woke up this morning" tradition—desperate odes (usually referring euphemistically to sex or death—or hangover-induced cravings) to hot coffee, sugar plums, jelly rolls, rolled oats . . .

In "Illegal Smile," John Prine, the whimsical master of the morning after, loses a staredown to a bowl of oatmeal. Then, in "Please Don't Bury Me," he doesn't even make it to the breakfast table—just wakes up, walks into the kitchen and dies. You'd think that'd be the shortest, most fatal blues morning ever—except there is a bootleg recording of Jimi Hendrix and Jim Morrison madly carving up the blues, and one of the tracks is creepily called "I Woke Up This Morning and Found Myself Dead," though the actual line is hard to make out amid their drunken incoherence.

There's even some debate about Morrison's most famous morning-after line—one of the most rousing yet nihilistic hair-of-the-dog hangover lyrics ever. Ray Manzarek—who became the de facto voice of the Doors after Jim found himself dead in a bathtub in Paris, just nine months after Jimi did the same in a hotel room in London—says that Morrison wrote the lyrics of "Roadhouse Blues" upon waking up from three weeks of drug-induced sleep, and the line is actually "Woke up this morning / And got myself a *beard*"—which at first sounds like a surreal joke, and then becomes downright scary.

What did it say on the bluesman's tombstone?

"I didn't wake up this morning."

PART NINE

BEYOND
THE VOLCANOES

IN WHICH OUR SUDDENLY SINGLE MAN WAKES UP IN LIMBO,
IS PAINFULLY CLEANSED, LEARNS TO WALK, CLIMBS A FEW
MOUNTAINS, GOES TO A BEER FESTIVAL, ALMOST DIES, IS BURIED
ALIVE AND DISCOVERS THAT TIME IS FLUID. APPEARANCES
BY SILENUS, MARTIN LUTHER AND A JET-SETTING TOMCAT.

He was drunk, he was sober, he had a hangover; all at once.
—MALCOLM LOWRY, *Under the Volcano*

I WAKE IN A PARALLEL WORLD—A BIZARRE, HELICAL, optimistic one—as if formed from the dreams of some infant suckling deity. An impossible architecture of temples, minarets, towers and domes rises from deep beneath the Earth. Fields and forest from the surrounding hills stretch over roofs like the rounded backs of dragons, so that the land and man-made dwellings seem one and the same. Pools, fountains, gardens and underground grottoes connect through stone tunnels, over earthen bridges—a labyrinth of steam, sunshine and crystals.

Horizontally, this place is just a hundred acres at the foot of the Austrian Alps—but laterally, it is infinite. The water flowing through the network of baths, and into the bathtubs in the rooms, comes from a profound primeval sea heated by the planet's core. Giant passenger balloons rise into the sky, like Icarus with a picnic basket.

The surrounding hills are quiet volcanoes, making the land abnormally fertile, volatile, magnetic: ideal for orange wine and spiritual connectivity. Here, you can sleep like a monk in a soft, candlelit room devoid of all technology: no phones, computers or electric light. In the volcanic pools, fountains and saunas of the spa, clothing and swimwear are strictly forbidden. The people are naked, radiant and beautiful. The air smells like lilacs. It sounds like flutes. I take a breath and try to be hopeful, though I feel like fucking hell.

The past long while has been dark and rough—like stumbling through some surreal and boozy postmodern blues song, sung from my destroyed apartment in Canada to a writers' festival in Croatia, through the foothills of the Alps and now into purgatory: I lost my girlfriend. I lost my money. I lost my luggage. I lost my hangover pills. I lost my way, and now I've ended up here: in this strange volcanic world, this creative concept of balance and health—a holistic village of wood, stone, terra-cotta and glass, and not a single straight line as far as a soul can see.

For centuries, of course, all manner of high-flying reprobates, rounders and romantics have crash-landed their hangovers on lesser hills in hopes of serenity, if not salvation. "Retreats to which they could return," as Clement Freud put it, "when their wings had been temporarily clipped by unrequited love or disaster in the card room."

But this place is worlds beyond that: a cutting-edge cosmic plane where respite and rejuvenation have reached their *nth* degree. Designed by Friedensreich Hundertwasser—a mythically humanistic artist and architect—as "the world's biggest inhabitable work of art," and built by the revolutionary developer Robert Rogner as "a conduit between untouched nature and hopeful humanity," Rogner Bad Blumau is a romantic, spiritual health resort where all manner of healing power has been made essentially inescapable.

According to the signage, the water in these pools rises from a million-year-old ocean "hermetically sealed off from outside influence through tectonic movements." Flowing from a depth of ten thousand feet at a temperature of 230 degrees Fahrenheit, it "strengthens the skin and the metabolism . . . supports the blood circulation and has curative effects on the whole body." Even just lounging at poolside, the goodness is sure to get to you, as these springs exist on "highly sensitive ground, where a number of power points, centers of vital energy and natural intelligence can be found."

And so, crashed upon these celestial shores, I am more than ready to give myself up—completely, holistically, infinitely—to the healing powers that be. I know that *detox* is no longer just shorthand for the sobering up of assholes. But I am also aware that things like "vital energy" and "natural intelligence" are often lost on souls like me—or, at least, we have trouble finding them. And so, even the first line of the brochure handed to me upon arrival reads like a sort of divine admonishment: "Some people never have an opportunity to change their ways, others have one but fail to recognize it."

"I recognize it," I say, floating in the Vulkania curative lake, pan flutes piped through underwater speakers. "I recognize it," I intone between sips of wheatgrass and blueberries, sitting beneath a banyan tree in the Garden of Four Elements. "I recognize it," I gasp, gushing sweat in an infrared sauna powered by the heart of dormant volcanoes. "I recognize it,' I whisper, lying naked in a cavern of salt brought by the truckload all the way from the Dead Sea, then molded into stalagmites with crystals in them. There are soft-synth sounds and a pulsing light. I am told to breathe deeply. Then a woman speaks my name. "I recognize it," I confess. And she leads me out of the Dead Sea salt.

HEIDI, WHO SPEAKS very little English, lays me down on a square white bed in a white square room. "This is body drainage treatment," she says, carefully pronouncing every word, then pulls back the towel and conveys to me the following: this treatment is new—so new, in fact, that I shall be her very first man. Until now, she has only drained the bodies of women. But after this, I will be clean. Cleaner than I've ever been. Is this alright with me?

I tell her that it is.

She turns up the Enya and flexes her fingers.

Not only are Heidi's hands strong and willful, digging into my toxin-laden muscles, breaking me down—but she has an actual "body draining" implement: a small, incredibly powerful, gel-smeared vacuum cleaner to press into my flesh, to siphon the detritus as she puts me to ruin. The mechanism hums and Heidi whispers into my ear, pressing down on top of me. It is invasive and overwhelming, painful and sensual, until Dead Sea water is pouring from my pores, out of my eyes—like my whole body has started to sob . . .

And as I purge at the hands of Heidi, poisons pulled from deep within, I am spun through a series of flashbacks, a toxic clockwork orange, sheet lightning suppressed between blackouts: breaking up with Laura; breaking down in Croatia; a disintegration through Slovenia; so many awful, drunken, shimmering moments, leading me to here.

Heidi moves down, down, down, until she hits my kidneys, like a wall of stone. "Kidneys," I gasp. The word, in English, means nothing to her, and I start to babble, about kidney beans, rabbit punches, California swimming pools . . . I am pummeled and drowning beneath Heidi's healing hands, down and down, until finally she is done.

"Now your beans are clean," she says. "Drink a lot of water." So that is what I do. After dragging this sad collection of liver, heart and brain over oceans, across borders, into this otherworldly sanctum of health—at the end of the most healing-focused day I've ever experienced—I sit in hot volcanic water, drinking cold volcanic water.

Amidst the recent mess of losing things—all my stuff and sense of equilibrium, the love of my life and vestiges of sanity—the only thing that's turned up again is my battered, brutalized luggage. Full of the books I was to sign, then sell, the suits I was to wear, it looks like it was dropped from ten thousand feet, somewhere between Canada and Croatia. And my hangover cure, or as close as I've come to it, has simply disappeared. By this I mean the case of pills themselves, not some sort of master recipe. I imagine a frustrated pharma-pirate or clueless scavenger trying to figure out what the hell he's got a hold of.

But still I have hope—of somehow surviving this recent messed-up binge and drawn-out toxic crash, even without my magic pills. This extreme barrage of detox and purging might, I imagine, force a sort of balance—sidetrack a kind of snowballing hangover/heartache for which I don't yet have a name. I feel sad and exhausted, but also vaguely optimistic. I look at the stars, take a last drink of water and go to bed beneath the volcano.

I WAKE IN molten lava, burning as it hardens. My head will not turn on my neck, my muscles are bound in red-hot stone, my blood is like boiling mud. And now, as I try to breathe, the bed is filling with too much gravity—a deadly black hole on the edge of this briefly hopeful universe.

It is an eon or an hour—in slow, painful increments, breaking through magma and immensely heavy air—before I make it to the edge of the bed. Getting up and getting dressed is beyond all possibility. Neither my knees nor elbows will bend; my head is half-stuck in the collapsing of the bed-star; my hands feel like steel, drawn from a blacksmith's forge. I lift them up to my compressing head and drag myself onto the floor.

It takes some focus, and rolling around, but it's almost certainly still morning (or at least early afternoon) by the time I'm somewhat dressed: a molten Tin Man in a Bad Bluman bathrobe, lurching out into the hallways of Utopia. Every burning, jangling step sends shock waves through my organs. That I'll be forced back into nakedness should I manage to reach the volcanic spas is just one of a thousand ironies. Staggering down the Geometric

Path of Perception and Healing, they leap and overwhelm me, like dogs upon my back.

This place is lousy with healers, helpers, swamis and soothers, all of whom I surely need. But what am I to tell them: My chi is toxic? Heidi tried to kill me? I am allergic to vital energy?

Instead, I ask the way to the saunas, to sweat the sickness out. But I'm just too goddamn beastly. Stumbling through all these golden couples—glistening, naked, holding hands, serene, shimmering in the mist—I can feel their shining eyes on my hideous form, this chewed-up, burned-out pariah dog who's been sleeping with the monkeys.

I find a corner to drop myself down, and then can't get back up. So here is where I'll stay, in this singular form of purgatory: boiling in volcanic heat from two miles below the Earth; surrounded by naked Austrian lovebirds; trapped in the holistic design of a supposedly humanist artist who wrote upon the walls above me:

THERE ARE NO EVILS IN NATURE.
THERE ARE ONLY EVILS IN MAN.

TOXIC SHOCK

Until very recently, the term *detox* had more negative, but arguably much more meaningful, connotations than it does now— suggesting an uncomfortable process of being weaned from the throes of addiction.

Now the Gwyneths of the world can put it in front of anything— water, air, cookies—and we'll pay top dollar for the simplest things on Earth, or subject ourselves to any number of complicated discomforts—ear candles, enemas, footbaths of flesh-eating fish—safeguarding against some unknowable cesspool of toxins into which we've all been plunged, hoping to come out fresher and skinnier.

According to most medical experts, however, our bodies are already equipped to deal with the peril of toxins—thanks to such antiquated phenomena as skin, lungs and liver. Says Edzard Ernst, professor of complementary medicine at the University of Exeter, "No improvements are needed or can be achieved by detox therapies. Proponents of alternative detox have never been able to demonstrate that their treatments actually decrease the level of any specific toxin in the body." Ernst calls current concepts of detox "not just wrong but also dangerous," by which he means people might overindulge in things that are bad for them, thinking all can be corrected with the right treatment.

To be fair, such ideas having not yet being proven in accepted scientific studies doesn't mean they're impossible. All this time looking at the research, and lack thereof, regarding hangover treatments has at least shown me that.

But in regards to detox fads, might there be another danger entirely? Some sort of massage and body drainage treatment, for example, that—when brought to bear on a deeply saturated system—might not calm and cleanse, but rather push harmful toxins *into* the rest of your messed-up body?

Sure, it's a rhetorical question—but I'm not the only one asking. Dean Irvine, for another, went to a spa after a night of heavy drinking and then wrote an article for CNN's website entitled "When Massages Go Bad": "About 12 hours after feeling relaxed and at peace with the world, I felt awful. I had a thumping headache, my whole body was sore and by turns I felt hot and sweaty or cold and shivery." Such symptoms, combined with his massage therapist's explanation that a massage detoxifies by "encouraging your body to get rid of various toxins from your blood stream via sweat and the lymph system," and Irvine strikes on the seemingly reasonable hypothesis that the deep-tissue massage "unleashed a load of toxins that had hitherto been locked away and not really bothered me."

Science writer and former registered massage therapist Paul Ingraham, however, believes that what might be going on is something even more counterproductive and extreme: that instead of merely liberating existing toxins, a particularly aggressive massage could actually *create* them in the body. In his article entitled "Poisoned by Massage," Ingraham suggests that such treatment can, in fact, create a mild form of rhabdomyolysis, which is essentially a toxic shock (often found in victims of car crashes and earthquakes) that happens when muscle is crushed and cellular guts spill into the blood. Depending on the severity, this can cause anything from flu-like symptoms to muscle fever to the sensation of being buried alive in molten lava.

All of this considered, I have to suggest that deep-tissue massage, no matter how "detoxifying," is not the best way to treat a deep-body hangover—and may reveal an Alpine Austrian spa as the ground zero of *Verschlimmbessern*: a veritable Bavarian temple to making things very much better by making them so much worse.

BEYOND THE VOLCANOES (AND INTO THE ALPS)

I extend my stay in this Dr. Seuss limbo until I am finally able to move again. Then I get behind the wheel and drive farther into the mountains.

My ultimate destination, and why I'm heading this way to begin with, is Oktoberfest—and not only that, but the Oktoberfest Hangover Hospital: a massive, *M.A.S.H.*-style, hangover-themed, tent-city hostel on the edge of Munich, where drunken revelers from all over the world party and retreat, drink and repeat, under the kitschy care of the Hangover Hospital helpers, who dress up as sexy doctors and nurses. For months now, I've been making plans with the organizer, who has offered me full access and my own press tent—as long as I can get myself there.

Oktoberfest is still a week away—and much has gone side-ways since I first booked this trip—but I am trying to stay on track. I have heard tales of an Old World mountaintop hangover remedy where they boil you in a coffin and bury you in hay. So that is where I'm headed. And it just so happens that, on the way to this particular mountaintop, is the Altaussee Kirchtag—an ancient annual drinking festival of which there are also many tales. I figure that's where I'll acquire the necessary hangover. All the inns in Altaussee, however, have long been filled, so I'll be staying one town over—in Grundlsee.

The drive should be only a couple of hours. But these are the Alps, my car is Croatian and bereft of GPS, and I do tend (even when not crawling out of purgatorial volcanoes) to choose the road less traveled. So it takes a while to get there.

When finally I arrive, my room is so ridiculously romantic it makes me physically ache: an open-concept suite descending in stages from kitchen to dining area, to Jacuzzi, to couches, to bed, to large French doors opening onto a stunning view of the lake—on the far shore, a castle, with turrets like folded wings. I open the bottle of complimentary champagne and try to think only of hangovers, and not lost girlfriends.

Beneath me is a spa that reaches onto the pier, so one can slip from the saunas into the cold, clear lake. But even more notable, regarding Alpine hangovers, is to whom this place belongs: the Seehotel is owned and operated by the Austrian manufacturers of the largest-selling energy drink of all time, so half the minibar contains booze and the other half Red Bull.

To avoid driving home from the drinking festival, I decide I'm going to walk—from the town and lake of Grundlsee to the town, lake and festival of Altaussee, just one mountain over.

"If you must walk, you must go around the mountain," says the hotel manager, giving me a map and pointing out the path. "It is something like two hours. Or maybe two and one-half . . ."

TWO AND A half hours later, I am deep in the woods, on a steep incline, and have been so for a long, long time. I drag myself up and around one more corner, then stop dead: I am teetering on top of a mountain, a foot from the edge of a cliff.

Three thousand feet below is the lake and town of Altaussee. I catch my breath, step back from the edge and see a nearby lookout tower. I climb it slowly and look around. I can see the town from which I've walked, about the same distance as the one ahead. I have, apparently, taken a wrong turn and gone *up* the mountain instead of *around* it. If only there were something that could give me wings . . .

After hours more of misguided trudging, and the ever-present edge of a cliff, the sun is going down on me. And though I can hear the sounds of a festival rising in the distance, I just can't get to it. I am slipping and stumbling through the woods, down slopes and cliffs, until finally, wild-eyed, cut-up and drenched in sweat, I emerge from the forest, at the base of the mountain, onto the lip of the lake—the *wrong* lip. I can see the town of Altaussee. I can hear the people singing—but on the far side of the lake.

I stare across the water, considering my options—until a growing, pent-up thirst for beer overrules the rest. I take off my clothes and put them in my pack. Then, naked but for the pack on my back, I wade into the dark lake and start to swim. From across the water come flashes of light and a deep, resounding howl.

FINALLY, SOME SIX hours after leaving Grundlsee, some ten hours from the purgatory of Bad Blumau, I crawl out of the dark water, put on my wet clothes, trudge across a field and come to a massive, glowing doorway. Inside is another world: golden-haired people by the hundreds, in lederhosen and dirndls, standing on massive banquet tables, crashing steins together, singing along to an oompah orchestra.

The air is thick with beer, sweat, hay, bratwurst, honey and pumpkin. And although constructed each year anew by the local volunteer firemen, this massive beer hall full of Austrians young and old feels timeless, heroic and insular: as if crassness and cell phones have been somehow made verboten.

The only incongruity amid all this Old World Bavarianism is the Red Bull tent outside the beer hall door, and the constant flow between the two.

A WHOLE LOT OF BULL

In regards to drinking halls, habits and hangovers, the most obvious turning points in Western history go something like this: the discovery of distillation, the Industrial Revolution, the disaster of Prohibition and the winged rise of Red Bull.

And as with each of those other phenomena, Red Bull hit with all sorts of hyperbole, as both a blessing and a curse—particularly in regards to hangovers. It would pick you up the morning after, and also "give you wings"; but mix it with booze the night before, and you got a whole new kind of cocktail: the Raging Bull, the Bulldozer, the Depth Charge, the Speedball, the Swift Kick to the Nuts . . .

In the mid-'80s, when it first appeared, Red Bull was seen as a harbinger of, and a revolution in, the way all-night party kids would party all night. Some of that was true, insofar as corner stores and gas stations seem mandated now to carry no fewer than 300 brands of energy drinks. And the dual purpose of vitalizing and *re*vitalizing is now marketed directly by many of them, with words like *rehab*, *repair* and *rebound* in lightning-bolt script across selected cans. But whether any of this has made drunkenness or its aftereffects fundamentally different is anything but clear, as evidenced by the spectrum of experts.

On the Hangover Heaven website, for example, our old friend Dr. Burke writes: "After treating thousands of hangovers (more than any other doctor in the world), I have noticed that people who drink more than 2 or 3 energy drinks in a night tend to have some of the worst hangovers imaginable. . . . When I go to do an in-room treatment and I see empty 12 packs of Red Bull outside their room, it is always a sign that I have my work cut out for me." Burke concedes, however, that "more data is definitely needed before we conclude how culpable the energy drinks and alcohol combo is in the increased number of deaths and hospital admissions."

Dr. Joris Verster, founder of the Alcohol Hangover Research Institute, and whom I met for coffee outside Amsterdam, has been trying to generate precisely such data. In light of recent studies and reports suggesting that alcoholic energy drinks are riskier than classic cocktails, he took on two main truisms: that when people mix booze with energy drinks, they consume more alcohol, and that energy drinks mask the feeling of intoxication. The results of his survey of two thousand Dutch students (which, it should be mentioned, was funded by Red Bull) contradicted both these claims.

It should also be mentioned that Red Bull has never officially promoted mixing energy drinks with alcohol—whereas several other companies have, at some point, done just that: such as Four Loko, Sparks and Tilt.

Then there's Purple, a product taking the current "detox" fad
to a whole new level of confusion. Marketed as a hangover-proof
energy drink-slash-cocktail mixer, Purple's slogan is "Detox as
you tox," and CEO Ted Farnsworth has been quoted as saying,
"You drink it with vodka until you can't walk, and if you have
a hangover, I will buy it back from you—including the vodka."
How such a post-peristalsis transaction could possibly take place,
however, is uncomfortable to imagine.

BEYOND THE VOLCANOES (THE HILLS ARE REVIVED)

In the end, it was one of those nights—shattering, incandescent,
when even into the deepest darkness you are flying too close to
the sun, tumbling with melted wings, then scooped back up by
some champagne supernova; one of those nights when you are a
lost beast crawling up from the Black Lagoon, don't even speak
the language—and then somehow it all just clicks, into a golden
moment, a fireball in the mind, and suddenly it's like the night is
yours: a hundred new friends sloshing ale into the air, drinking
to your health and the magnificence of your death, while the gal
with whom you were dancing is now conducting the oompah
orchestra; one of those nights that are incalculably dangerous,
yet it is impossible for you to get hurt; one of those nights you
must never, ever speak of—as, in the end, it isn't yours.

Yours is the morning after, when you wake up bereft and hollow, get behind the wheel and float up, up, up, into the Carinth-
ian Alps.

I have seen some awesome things before (and also *The Sound
of Music*), but nothing has prepared me for the jagged, luscious
immensity of the Alps: expanding greens and blues, like being
inside a rough-hewn water molecule, like looking at Earth from
the Moon. And I've gone over so many mountains by now that

why start going around them? So I take them straight on, pounding Red Bulls, overwhelmed by the breadth of the world and the sudden truth of cliffs.

There are no guardrails, no safeguards, nothing to slow the fall. I go up and up, and then right into the clouds, so that now the world is gone, and when I take the next turn and start on the new descent, I can't be sure if I'm driving or falling. I move faster, and faster, then out of the clouds, somehow still on the road, on the side of the mountain, on this lost planet, spinning in space; then up again, up and up and up, until finally I realize: *This road is going nowhere.*

Once made for desperate loggers, or daredevil snowmobilers, or who the hell knows, the road has all but disappeared: so steep it's almost vertical, so narrow—a cliff rising on the right, a mile-high drop on the left—that trying to turn around is probably a fatal impossibility. And climbing any higher will only make it worse. Even as I realize this and start to panic, my four-cylinder, four-speed, standard-shift Croatian car slows right down . . . and is about to stall. So I hit the clutch and the brakes.

And now I am totally fucked.

The moment I release this clutch, I will hurtle backward down the mountain. To hurtle forward, I would have to turn this car around. But three feet from my wheels is the edge of the cliff, and then nothing but air for miles. My legs are shaking, and if my feet falter, I am dead. I reach over slowly and pull the emergency brake. And now my whole body is vibrating, adrenaline blurring my blood. I can feel my heart like a sledgehammer, swinging to get out of my chest. Fuck, I don't want to die. I clear my throat and say it out loud.

"Fuck, I don't want to die."

I picture all the things I'd have to do so quickly to somehow stay alive: release the hand brake, the foot brake and the clutch while gunning the engine and turning toward the edge of the cliff; hit the edge without going over; then stop and do it all again, this time in reverse, again and again, until the teetering inches become a car turned around and sliding back down—instead of a crumpled mess a mile below, somewhere in the Alps.

In this thumping, still moment, I can see the discovery—maybe years from now: a skeleton in my clothes, and the battered suitcase in my trunk, a mystery to the investigators, full of unsold books and unworn shirts. I grip the wheel with my left hand, the hand brake with my right, flex my toes, and take a breath . . .

HOURS LATER—THOUGH my baffled watch says only minutes—my little Croatian car is pointed *down* the mountain, and I am soaked in sweat. I let the car descend slowly, from dirt, to gravel, to rock, to pavement, then I pull to the side of the road. I get out of the car and vomit. I lie on my back, staring at the spinning sky, and stay like that for a long, long time.

When finally I arrive at the remote resort of Almdorf Seinerzeit, it is precisely what I hoped for: the kind of place, atop an Alpine peak, where a reclusive mythic guru, a wealthy warrior-monk in training or a recuperating spy might finally be discovered.

It is also popular with nature-loving newlyweds, who can chop their own firewood to heat the wooden hot tub on the deck of their two-story log cabin, from which they sip a glass of local sparkling wine and look out on the world below. But I am just here to be boiled in a coffin, then buried in a pile of hay.

WHERE, PRECISELY, THE idea of the *Kräuter-Heubad*, or herbal hay bath, comes from is difficult to say. "It is from here," says the woman with night-black hair and green glowing eyes, but her hand gestures out at the twilight expanse of mountains, slopes and valleys beneath us. Her hair is tied in coils that whisper as she opens the door and guides me down the steep stone steps into a candlelit, stone-walled cellar.

Before a roaring fire sits a wooden, body-sized box, with a hinged lid, full of steaming water. I undress and climb in. The lid, when closed, goes only to my shoulders, and I can see before me a giant black cauldron, hanging on chains in the fiery hearth, as she moves toward the fire. Brewing in the cauldron are a hundred healing herbs, cut by scythe from the surrounding slopes; she scoops near-boiling water from the cauldron into the coffin through an opening at my feet. And I just lie there in the flame-flickering dark—so strangely, suddenly restful, for the first time in what feels like forever. I close my eyes and let my body drift.

When it's time to emerge from the water, she guides me out and onto a bed of hay. She covers me with a sheet, and then a lot more hay. The weight is both heavy and blissful, and a memory takes hold of my body: my baby boy asleep on my bare chest, his warm bare skin against mine as we breathe. A myriad of pain radiates, then dissipates, out from around this place on my chest, where it helps me to remember. I let myself be buried.

When she finally digs me out, the raven-haired, green-eyed woman pours oil on my skin for a full-body massage.

"The last time this happened," I tell her, "I very nearly died."

"Then this will be gentle," she says.

When her fingers lift, I feel like I am sleeping, but standing and wrapped in a scarlet robe. Then I rise, step by step, up the stone staircase, from the depths of the Earth and into the mountain air. The sky is a brilliant blue, full of golden stars. I eat my dinner beneath them, drinking orange wine. It is my last night among the volcanoes, and I feel like a resurrected beast—just alive enough to go back down to where the townsfolk live.

DRUNKEN SILENUS

In Ovid's *Metamorphoses*, there is a sort of impromptu victory parade where Dionysus, returning from some high-flown adventure, rides through the streets of town on a chariot pulled by wildcats . . . and followed by Silenus:

> *The lynxes draw his car, with bright reins harnessed,*
> *Satyrs, Bacchantes, follow, and Silenus,*
> *The wobbling old drunkard, totters after,*
> *Either on foot, with a stick to help him hobble,*
> *As shaky on three legs as two, or bouncing*
> *Out of the saddle on his wretched burro.*

In the pantheon of chiseled, superpowered Olympic gods, Silenus is by far the least impressive: a pudgy, hapless Bilious to the wine god's Bibious; a sort of ancient mythical Galifianakis. He is what happens if you imbibe like a god—even if you *are* a god—but are not Dionysus. He is the bearded, overweight Jim Morrison, endlessly pissing in a French bathtub. He is leisure-suited Elvis shooting at TVs in Graceland for all eternity. He is the hungover counterweight to Dionysian invincibility. And staring at his likeness in a masterpiece in the Munich Museum of Art, I can't help but feel a wretched sort of kinship.

The Drunken Silenus, painted by Peter Paul Rubens circa 1618, is remarkable in a few ways, beyond being the work of a master. Throughout most of history, and despite a pretty much constant litany of debauchery and debasement, Silenus managed to remain remarkably good-natured—laughing with self-deprecation even while falling off a donkey, arms open rather than trying to break his fall. But here, he is caught by Rubens in what seems like personal distress. And the artist has made it even more so by painting his own face, befuddled and downcast, on the body of this bloated

deity, stumbling through a bacchanalia of Rubens's closest con-
fidants—depicted here as nymphs and satyrs. They observe his
attempt at progress with gentle, yet warranted ridicule.

BEYOND THE VOLCANOES (AN ATTEMPT AT PROGRESS)

Things have not gone well since I came down out of the hills. I'd
hoped to get back on track in my quest for present-day cures.
But the makers of Kaahée, an Austrian hangover remedy, have
disappeared since I talked to them from Canada. Same with the
German remedy Cure-X; it seems to no longer exist.

And now, the whole reason I came all this way—to Munich in
the autumn—appears to be dead in the water, or at least drowning in a keg of pilsner.

"It's pretty complicated," says Guliano Giacovazzi, looking deep into his stein as we sit in the beer hall where Hitler apparently used to hang out. Giacovazzi is an Italian–South African party promoter and the head of Hangover Hospital—or at least he *was* before some sort of coup took place and his business partners absconded with the tents, banners and sexy triage costumes. Or something like that.

There's certainly more to all of this, but what matters most for our purposes is that the Hangover Hospital will not be going ahead—at least, not with Giacovazzi at the helm. I've returned my rental car and no longer have a place to stay, and to find one in Munich at the start of Oktoberfest is practically impossible.

"A buddy of mine said you could crash on his couch," offers Giacovazzi, trying his best to be helpful. "Or you could check out Stoketoberfest."

An extreme campground experience for full-on party kids, Stoketoberfest is supposed to be much like Hangover Hospital, just without the wholly relevant theme. And Giacovazzi believes he can score me a tent.

"Sounds good," I shrug. "A chance for more adventure." But the truth is, I am pretty much done, sick to all hell of adventure. The last place I want to be right now is Munich for Oktoberfest, with no particular spot to lay my messed-up head. And the last place *within* that place I want to be is this stupid, famous, overcrowded beer hall where the teetotaling future-Führer apparently pretended to drink. There are signs in the bathroom reminding foreigners, through pictographs, that men in Bavaria should pee sitting down.

When the beer hall closes, I drag my luggage through the streets, piss in a fountain in a public square—half-hoping some German cops will show up and tell me to sit down to do it. Then

I throw my stupid, busted-up suitcase into the fountain and leave it there, half-submerged like a body in a crime scene that will never make any sense.

THE KATERS OF THE WORLD

In my twenties, I lived all over Mexico. And up until starting this book, that's probably where I learned the most about hangovers. I made friends with a doctor in a small coastal town, and for the cost of a half dozen *cervezas*, he would shoot a syringe full of vitamin B_{12} and Toradol (salvaged from the latest earthquake-relief effort) into the veins on my arm. And it did tend to work—at least enough to get me drinking again. Like most Mexicans, Dr. Vaso called a hangover *cruda*, which means "rawness." The way he said it made it the most perfect word for the morning after. Until I moved to Spain.

There, the word is *resaca*—meaning, what is left behind when the tide recedes: flotsam and jetsam; the detritus of an undertow or ocean storm. What in the world, other than maybe *hangover*, could work so poetically, on so many levels?

That very question is why I am here, in this white-walled, sunlit room, as the sound of waterfalls in rainforests beats softly from invisible speakers. This futuristic German building feels more like a health spa than the hub of a worldwide ad agency. Tricks of light, elaborate windows and wall projections blend the outside hanging gardens with streamlined corridors and conference rooms. On the table in front of me is a small, black, shaggy square. It feels like I never left limbo.

"That's the last remaining furry one," says Martin Breuer of Havas International, pointing to the book on the table. "It's a prototype. You're welcome to take a look."

Die 10 Berühmtesten Kater der Welt (The Ten Most Famous Hangovers of the World) was the result of an advertising initiative created by Breuer and his team for the product Thomapyrin. Available in Germany and Austria, Thomapyrin is a mix of ASA (Aspirin), paracetamol (Tylenol) and caffeine. It is equivalent to Excedrin in the United States and Anadin in the UK and is recommended for migraines.

But like any effective pain reliever, it also tends to be sought out when and where hangovers are rampant (sales spike at New Year's, for instance). So upon acquiring the account, Breuer wanted to harness that incentive in a creative way, so that people would think of Thomapyrin whenever they had a hangover.

In Germany a hangover is called a *Kater*, or tomcat.

"The idea became to embrace the *Kater*," says Breuer, as I open the book, "and then ask the question of what it really is, in a universal way. Does it have the same feelings, the same connotations, in different parts of the world?"

The original plan was to have international artists render their homeland's hangovers, along with some local remedy. But when Dennis Schuster, a renowned German street artist, offered up his strung-out cat, drawn in an urgent, yet timeless, woodcut style, it just fit the idea so well—a literal *Kater* subjected to ten of the world's euphemistic, yet all-too-real ones.

I've studied these pages before, online. I love the images, and the whole feel of the book. But then, when I had the accompanying text translated, I found it somewhat baffling: very few of the definitions, and almost none of the suggested cures—though certainly familiar—corresponded at all to the research I'd done.

But Breuer just meets this with a shrug: "It is meant to be fun—to make you feel something and think of something new. For instance, most of the campaigns we are known for, they are colorful. But this one is not. The hangover is not colorful. It is a bit dark. These images, yes, they are a little bit fun—because drinking is fun. But they are also dark. Everyone likes the evening before, but nobody likes the morning after. You get that from this cat."

So here is that cat, with thanks to Dennis Schuster. The accompanying (sometimes puzzling) text has been translated from the German, then edited or summarized for space:

Kater ['kaːtə]

1. KATER (GERMANY)

Origin: A nineteenth-century beer called *Kater* ["Tomcat" in English], favored by German students.
Remedy: Fresh Owl Eggs. [As per usual, this "cure" is accredited to Pliny. But it also astutely mentions that owl eggs contain cysteine.]

Tømmermænd [tœmɐmɛnd] → Zimmermänner

2. TØMMERMÆND (DENMARK)

Origin: *Tømmermænd* means "carpenters." Danish carpenters toasted each new roof beam with another drink.
Remedy: Fireplace ash in a glass of milk.

Hangover ['hæŋəʊvə] → Ruм hängen

3. HANGOVER (USA)

Origin: A baffling explanation involving American settlers, land and drinking competitions, with the "overhang" of real estate as the ultimate prize.
Remedy: Powdered swallow beaks mixed into a paste with myrrh, then eaten.

Resaca [rre'saka] → Meeresbrandung

4. RESACA (SPAIN)

Origin: A convolution of everyday, placid surf with some abstract notion of seaside diners waving incorrectly at Spanish waiters.
Remedy: Rubbing your armpits with half a lemon.

Babalas ['babalas] → Der Morgen nach dem Tag davor

5. BABALAS (SOUTH AFRICA)

Origin: Apparently, this is an ancient Zulu word meaning, quite simply, "The morning after the day before."
Remedy: Pickled sheep's eyes in tomato juice.

Gueule de bois [gœl də bwa] → Hölzernes Maul

6. GUEULE DE BOIS (FRANCE)

Origin: Translated here as "wooden muzzle," which refers not only to the sensation of waking up with a mouth full of wood and sawdust, but also to the custom of aging Cognac in wooden barrels.
Remedy: Diced garlic in red wine.

Suri [ˈsuːrɪ] → Schwindel

7. SURI (BAVARIA)

Origin: Possibly a slurred way of saying "sorry" . . . ?
Remedy: Two egg yolks and a shot of Cognac in a bowl of noodle soup.

Baksmälla [ˈbɔːaksmɛla] → Rückschlag

8. BAKSMÄLLA (SWEDEN)

Origin: [The one given here is complicated, involving manure and tabletops. But really, if the translation is right, it needs no explanation.]
Translation: Backlash.
Remedy: A classic mix of eel and almonds.

Pochmeliye [pac'mɛljə] → Pochen im Kopf

9. POCHMELIYE (RUSSIA)

Origin: Translation: Drunk stupor
and dizziness.
Remedy: Dissolve a cigarette in a
good amount of coffee. Pepper and
salt the concoction, look at it with
distaste and swear that you will
never, ever drink again.

Krapula ['krɑpulɑ] → Krabbe

10. KRAPULA (FINLAND)

Origin: [This apparently means
"crab," rather than "crap." But trying
to decipher the reason provided is
unreasonably exhausting. So I've come
up with a solution.] Translation: Crap.
Remedy: Stick 13 pins with black
heads in the corks of the bottles that
have caused the terror.

Hundreds of copies of this little black book—without the
furry cover but still quite beautifully made—were handed out during a boozy all-night German art festival, and the response was
overwhelming. It won an award for design, and images from the
book are still widely shared, illegally reproduced and celebrated.

And despite some consternation in regards to the accompanying text, I'm really a fan of this little black book; in fact, I'm
impressed that it even exists.

I'm so used to companies doing all they can to disassociate their products from the mention of hangovers (even many that were developed precisely with the morning after in mind)—or regional laws prohibiting remedial claims—that Breuer being able to do this, successfully, with no fallout or backlash, seems to belie all sorts of industry truisms.

But Breuer seems barely aware of it "Where is the conflict?" he says, as if genuinely puzzled. "If the product helps with hangovers, why not say it? After all, the most common thing people take is not sheep eyes and tomato sauce, or any of these other things. The most common thing is Aspirin. But *this* product is better."

He picks up the little black book and turns to the last page, translating aloud: "And by the way, when you want to get rid of the hangover *faster*, Thomapyrin helps. Just wanted to have mentioned that."

He puts it down. "Nice and simple, right?"

BEYOND THE VOLCANOES (WITH NOWHERE LEFT TO GO)

I have been trying to sleep in a campground full of impossibly intoxicated young alco-tourists as they vomit and pass out in all the wrong pup tents. I have worn lederhosen and glugged from a giant stein in a mammoth beer tent while Arnold Schwarzenegger conducted the oompah orchestra. At this point in my travels, my book, my life, I'm pretty damn sure there's nothing more I need to learn from Oktoberfest. And yet, my flight back to Canada is still a few days away.

Other than to see my boy, I am dreading it—going back to where the bridges are burning. But neither do I want to be here—or anywhere else, really. My mind and body remain volcanic, I have no paper money left and a decimated credit card, and any new idea seems as ludicrous as the last. I feel like old Silenus, soft and sick with drink, stumbling haphazardly across the Earth, careening down mountains, pissing in fountains, passing out amidst nymphs and satyrs in puddles of tears and Bavarian beer. I need it all to end.

I get on a bus with a handful of coins and ask the driver to take me out of the city, as far as he can. He picks out a few of the larger coins, drives for an hour or so, then lets me off, with a pleasant nod, in a small town whose name I'll never know.

With the remaining coins I buy a bottle of wine, a large piece of sausage, a loaf of bread and some Aspirin, then walk farther down the road, out of the town, and find a path into the forest. I walk and walk, and as it gets dark, I leave the path and walk some more.

ASPIRIN OR SORROW

The first pure, stable sample of acetylsalicylic acid was created in 1897. Until that time, there wasn't much that could kill pain without also getting you high. And it might be more than coincidence that Aspirin and the hangover came into our lexicon—and the popular consciousness—at the very same time.

Whereas it takes a lot of digging to find hangovers in literature before the final century of the last millennium, since then—and over the past hundred years—they have become one of the most pervasive subjects in all of literature. Just as most authors who take storytelling seriously eventually try their hand at rendering a love scene, or a death scene, or one in which their protagonist finally returns home, so too there is the "hangover scene." In fact, contemporary narrative is so rife—not to mention, bleary, tremulous, red-eyed and bilious—with them, that to compile a list of hangovers in modern literature is a task that might never end.

What a kiss is to love, a shroud is to death and a childhood bedroom is to coming home, Aspirin is to the hangover—a tactile, symbolic shorthand. But even that single trope, once you start digging, becomes a goddamn chasm. There are enough literary Aspirin moments to fill another book. So here are just a few.

In Hemingway's 1927 classic *A Farewell to Arms,* we find love, death, coming home and the chomping of Aspirin, all together in a single scene:

> "It's Austrian cognac," he said. "Seven stars. It's all they captured on San Gabriele."
> "Were you up there?"
> "No. I haven't been anywhere. I've been here all the time operating. Look, baby, this is your old toothbrushing glass. I kept it all the time to remind me of you."
> "To remind you to brush your teeth?"
> "No. I have my own too. I kept this to remind me of you trying to brush away the Villa Rossa from your teeth in the morning, swearing and eating Aspirin and cursing harlots. Every time I see that glass I think of you trying to clean your conscience with a toothbrush."

P.J. Wodehouse's Jeeves and Wooster stories, written and published between 1918 and 1975, pretty much span the first sixty years of literary hangovers. And there are many moments when Aspirin is required. In *Ring for Jeeves*, Wooster needs help sorting out what happened the night before.

> "Jeeves," he said, "I hardly know how to begin. Have you an Aspirin about you?"
> "Certainly, m'lord. I have just been taking one myself."
> He produced a small tin box, and held it out.
> "Thank you, Jeeves. Don't slam the lid."
> "No, m'lord."

In *Under the Volcano,* Malcolm Lowry's poorly disguised autobiographical novel of debauchery and heartbreak, a British consul staggers about Oaxaca like a hungover pariah dog, bereft of love, sanity and anything that might soothe—except for booze and Aspirin—until even those won't work anymore.

> [H]e was like a man who gets up half stupefied with liquor at dawn, chattering, "Jesus this is the kind of fellow I am, Ugh! Ugh!" to see his wife off by the early bus, though it is too late, and there is the note on the breakfast table. "Forgive me for being hysterical yesterday, such an outburst was certainly not excused on any grounds of your having hurt me, don't forget to bring in the milk," beneath which he finds written, almost as an afterthought: "Darling, we can't go on like this, it's too awful, I'm leaving—" and who . . . remembers incongruously he told the barman at too great length last night how somebody's house burned down—and why is the barman's name Sherlock? an unforgettable name!—and having a glass of port and water and three Aspirin, which make him sick, reflects that he has five hours before the pubs open . . .

Now we know that acetylsalicylic acid can cause gastric distress, internal bleeding and ulcers when mixed with too much alcohol—as can ibuprofen (Advil). And even small amounts of acetaminophen (Tylenol), when mixed with booze, can kill a heavy drinker by shutting down the liver altogether. And so it seems all our placid painkillers, meant to help without getting us high, might still destroy us in the end.

And finally, of course, none of this will matter.

As the late low priest Bukowski once wrote:

the dead do not need
aspirin or
sorrow,
I suppose.

but they might need
rain.

PART TEN

WHEN LIZARDS DRINK FROM YOUR EYES

IN WHICH OUR MAN, UNBOUND BY SPACE AND TIME,
TRAVELS INTO THE PAST, ATTENDS A BACHELOR PARTY,
CURES THE HICCUPS, CRAWLS ACROSS THE DESERT AND WATCHES
THE HANGOVER FOR THE FIRST AND SECOND TIMES.
APPEARANCES BY JOHN DILLINGER, BORIS YELTSIN,
RA THE SUN GOD AND A VAN FULL OF BRIDESMAIDS.

Drunkenness is nothing but voluntary madness.
—SENECA

I T IS ALMOST A DECADE AGO NOW, AND I AM YET TO officially embark down the long, treacherous roads of fatherhood and hangover research. Still relatively young and undaunted, I am on my way from Las Vegas to a bachelor party in Tucson—with my pregnant girlfriend and two sprained ankles.

The first one I sprained a week ago, in the rain, at a music festival on a wolf sanctuary, as Paul Quarrington, who has taught me almost as much as my father about writing, living and drinking, finished his final set, singing between breaths from an oxygen tank. My closest friend for the decade I've lived in Toronto, he was diagnosed with stage IV cancer the same week I learned I'd be a dad.

The storm was torrential. Paul was dying too fast, and we'd run out of whisky in the festival tent. But as I went out through the darkness to get a bottle from our cabin, my foot found the kind of sinkhole that breaks the leg of your horse as you fly over its head and hit the ground already crying.

The other was just a few days ago, on a steep trail into a canyon behind a roadside motel. I don't even know why I was going down there—clutching the goat-head cane I'd bought in Vegas. I stumbled, my bad ankle throwing me onto the other one—and that one crumbled too. So there I was, partway down a motel canyon with two torn ankles and a girlfriend with child at the top, watching me sob into the hot desert sand.

"Look at it this way," she said when I made it back up. "You've evened out your limp."

A FEW HUNDRED miles later, she drops me off in Tucson, at the hotel where John Dillinger, under an alias, hid out in his final days. It has that kind of feel to it—like the next slow-motion shootout or well-dressed bar brawl will be starting any minute. So, in the spirit of going on the lam, or to an ill-conceived bachelor party, I will slightly alter the names and appearances of those in attendance.

Staying at the hotel along with me are El Diablo, Thomas Crown and Ted. And while we're all drinking too much with the Impending Groom, the mother of my child-to-be will continue on to a resort near the Mexican border where her best friend, Rapunzel, is preparing to get married.

El Diablo is Rapunzel's younger brother—a smart, charming Van Wilder type who is known to get carried away. He could be played by some of the best Ryans in Hollywood. He is sharing a room with Thomas Crown, whom I first met at his own wedding, when he married El Diablo's other sister. Mr. Crown is a good-natured and quirky British CEO of a high-profile multinational; picture Hugh Laurie and Richard E. Grant, but with bespoke spectacles instead of all those bulgy eyes. In the room between theirs and mine is Ted—a family friend so simultaneously nondescript and deadly smart, he'd make a perfect spy.

The four of us are the only non-Americans invited for the stag, and I like and trust them all—in an optimistic, citizen-of-the-world kind of way.

At the hotel, we raise a glass to the groom, then take a cab to the suburban bungalow of one of his brothers for the beginning of the bachelor party. The front room is full of tequila bottles meant to be emptied in particular ways, scantily clad professionals and a bunch of guys with same-day haircuts and experimental cologne.

A few hours later, when the limo arrives to take us to a neighborhood gentleman's club, not everyone is able to go. The Impending Groom is passed out among quilted duvets while others are still vomiting in the expansive backyard. Mr. Crown and El Diablo decide to stay behind to care for the casualties while Ted and I venture out.

By the time we meet up again at our hotel in Tucson, stuff has gone down that I really shouldn't speak of—as well as a lot more tequila. And now this hotel bar is hopping—full of lovely gals. But despite his youthful charm and cevilishness, El Diablo appears to be floundering. So Ted and Mr. Crown appoint me his wingman before calling it a night.

And then I get the hiccups.

To fully understand what happens next, one would need to know my relationship to hiccups—which I think may verge on the pathological. In my own personal hell, there would be no booze: just hangovers, heartbreak and hiccups. I've loathed the latter for so long now, I've tried a hundred remedies. And the only one I've found that works is this: a tablespoon of sugar, soaked in vinegar, taken in a single gulp. Your throat does a sort of spasm, then voilà—and hopefully without vomit.

So that is how I find myself—a drunken, hiccupping, crippled wingman—ordering shots of vinegar and sugar at 2 a.m. in Tucson. It takes some explanation, between *hics* and *cups*, to convince the barkeep, but finally he pours me one—and then another. And that's when things get fuzzy . . .

ELPENOR ON THE ROOF

"'Tis the wine that leads me on," says Homer's Odysseus, deep undercover on a questionable mission. "The wild wine that sets the wisest man to sing at the top of his lungs, laugh like a fool . . . it even tempts him to blurt out stories better never told." Those, of course, are the tales not of bravery and heroism, but mishap and madness—like Elpenor on the roof.

The youngest of Odysseus's men, he was there when they won the Trojan War, defeated the Cyclops by getting him drunk, and journeyed for years across the "wine-dark" seas. But the beginning of the end for Elpenor comes in book 10 of *The Odyssey*, when they arrive on the island of Aeaea—where the impossibly sexy Circe lives with her nymphs.

She serves the soldiers an elixir (now thought to be a potent mix of mead, beer and wine) that turns them into pigs. In planning their rescue, Odysseus gets hold of a special herb (eventually identified by literary botanists as something similar to mandrake) to protect him from the drink. Then he overpowers Circe and forces her to change his soldiers back into men. After freeing them, however, he falls under a simpler spell: that of wine and Circe. Once seduced and satisfied, it is the placating feel of afterglow, and intoxicated abandon, that finally enslaves Odysseus: the idea that he could simply let it all go, so far away from everything, and just keep drinking, on into oblivion.

The booze has always been there, in every story, as long as we have lived them: a gift from the gods, a nefarious trap, a truth-telling serum, the devil's brew, fundamental medicine, insipid poison, pure depressant, inspiration liquefied, liberation intrinsic, a monkey on your back, a fiasco, hellfire, the signature of civilization, sunshine held together by water, darkness in a bottle, the night before, the morning after. We make it, celebrate it, question it, own it, kill it, damn it—then do it all again. It makes us, celebrates us, questions us, owns us, kills us, damns us. But if we're lucky, we find a way to stop.

It took a full year for Odysseus's men to shake his head all the way clear and make him want to go home—to Ithaca and sanity, and their families and children.

But then, on the night before leaving this island of booze and nymphs, young Elpenor got way too drunk and woke up on the sunlit castle roof to the sound of the anchor lifting. Panicked, disoriented and about to miss the boat, he slipped, or stumbled, and fell to his death—one of the most random, yet epic missteps in all of hangover history.

Not only did it alter the fate of Odysseus—who then felt compelled, in hopes of saving Elpenor's careless soul, to sail into the underworld of Hades instead of straight back home—but even today, banged-up, clueless people get diagnosed with Elpenor's syndrome: a kind of mania brought on by sleep disturbance or a late-drunk/early-hangover state so intense as to cause disorientation, delusion and drastically dangerous behavior.

WHEN LIZARDS DRINK FROM YOUR EYES (AND THE PHONE KEEPS ON RINGING)

I wake, eventually, to the ringing and ringing.

"Hello."

"Hello? What happened last night?"

"Stupid stuff. But nothing too bad. Why?"

"Because the guys who were supposed to be your ride got kicked out of the hotel last night and drove down here instead. And you are still in bed."

"What?"

"Exactly."

I go over to talk to Ted.

Apparently, sometime after passing out, El Diablo got up to take a piss. Then, sometime after that, there came a pounding at the door. Mr. Crown opened it to find the hotel manager and two policemen: one holding handcuffs, the other a cell phone. They were looking for the person pictured in this video, and turned the phone so that Mr. Crown could see it: shot from below, a young man, naked on a balcony, peeing on the heads of people partying in the courtyard below. He looked like a Hollywood Ryan.

The cops could see that this British man in briefs wasn't their latest Dillinger. "But this *is* the balcony," said the manager, stepping into the room.

"And yet," said Mr. Crown, more surprised than anyone, "it appears no one else is here."

The cops looked around—in the bathroom, on the infamous balcony—but it was the manager who spotted, finally, the bright green foot sticking out from under Mr. Crown's bed. They pulled on it, and out came El Diablo—naked but for the green socks, and still in the fetal position—like some overgrown newborn leprechaun.

"Damn," I say.

"Yes," says Ted. "At least they didn't arrest him."

So, now plans have changed, and another of the groom's brothers—who didn't come out last night because he no longer drinks—is going to pick us up and take us to the resort, sixty miles south, to attend the rehearsal dinner. But he won't be arriving for a couple of hours; first, he has to collect the Impending Groom, and some other things for the wedding. So I ask Ted to wake me in time, and I go back to bed for a bit.

OF COURSE, ALL of that is just the *before*. What matters is the *after*—the transition into which could well have been written by Alex Shakar in a passage from *Luminarium*:

He dozed off into a dreamless oblivion, for what seemed
like seconds but was in fact hours, and awoke hungover,
the inner surface of his skull pulsing like a single giant
nerve being chewed by some ruminant animal.

I gasp awake and feel the teeth chewing on my skull. The
world has somehow turned: painful, burning and sideways. It
requires clinging to the walls just so I can leave. Outside, the
air is unimaginable, on fire. It is like walking, or rather double
limping, into the mouth of a dragon. I pray it might incinerate
the animal in my brain. And also the one in my gut. The gnash-
ing now is everywhere, inside and out, as I climb, sideways, into
the idling SUV.

In the front passenger seat sits the Impending Groom—one
of his hundred brothers beside him. "Chip!" says the brother,
banging the wheel in time to Limp Bizkit, and he hits the gas as
Ted fumbles for a seat belt.

Chip! still has some stops to make before we hit the highway. Closest is the brewery, to pick up some kegs of beer. So that
is where I vomit first—in a bathroom so thick with the smell
of hops that, when I gasp for breath, it is like sucking stout. I
throw up again at the next stop, where we pick up tablecloths
and napkins; then out the open car door, just before the highway;
then once again when we've picked up speed, right out the open
window. It streaks behind us like a glowing oil slick, and Chip!
honks the horn as the guys all cheer. "What the hell did you drink
last night?" he says. But there is no way I can answer. The tail of
the spinning, gnawing ruminant animal is flailing in my throat, a
furry claw reaching through my mouth.

"A whole lot of tequila," says the Impending Groom, who
had promptly thrown up, then passed out peacefully an eighth of
the way through the night.

"And apparently," says Ted, in perfect deadpan, "a whole lot of vinegar too." The various burrowing rodents inside me start to writhe and spasm, and even as I try to call out to stop the car, we are already slowing down and pulling over.

"Fuck," says Chip! as I open the door to scream little bits of vinegar-and-tequila-soaked beasts into the burning dessert. Then I wipe my mouth and pull the door closed. It feels like the air conditioning is no longer on.

"Fuck," says Chip! again. I attempt an apology, slumped and roasting against the door.

"Believe it or not," Ted says calmly, "*you* are not the problem."

"Fuck, fuck, fuck!" says Chip!

"It appears we're out of gas."

"Fuck, fuck, fuck, fuck, *fuck!*" says Chip! "I *knew* there was one more stop to make."

All is silent for a moment, and then the heat becomes unbearable. We get out of the truck, which I hope at least might cast some shade for curling up to die. But the sun, by now, is directly overhead—a terrible, merciless god.

"Is there any water?" I try to ask—knowing if there was, I couldn't keep it down.

"No," says Ted. "But there's forty gallons of beer." So they tap a keg as I limp into the desert. But my legs can't hold me, so I fall to my knees and crawl across the blinding, sizzling earth, searching for a bit of shade.

I come upon a cactus and collapse beside it, looking up at an arm of spikes that cuts across the flaming sky—so many perfect swords, wielded somehow in an effort to protect me, to deflect just a hundred of the sun god's billion ruthless rays. Ra. *Rahhh!* I feel the animals twisting as they boil inside. It seems suddenly wondrous that I could die before Paul, beneath this brave and hopeful cactus. But then comes the thought: I would never meet my son—my boy, the blessed light without this heat—all because of the goddamn hiccups. And that's when I pass out.

YOU TUMBLE FROM dreams of deserts and demons into semi-consciousness. Your mouth is full of sand. A voice is calling from far away, as if back in that blurry desert. It's begging you for water. You try to move, but can't.

A NEW KIND of beastly scratching comes, not from within but without. *Here be dragons*, I think. I open my eyes and the scratching stops, for barely an instant, then something like whips or snakes lash across the world, faster than I can blink. My eyeballs need to get away. But my brain is stuck in reptilian fury. *I am the Lizard King. I can't do anything. My kingdom for a bathtub. Just let me drown in Paris water. Anything but this. They are drinking from my eyes! They are drinking from my eyes!*

Then, in an instant, the lizards are gone—and so, too, the sun. A shadow falls across me. A click, a beep and then the voice of Ra. "Holy man," it says. "Now I don't know which to screen at the rehearsal dinner: him peeing on people, or you babbling in the desert."

"You are a spy, not a sun god," I try to say, but it comes out like dragon's breath.

OF HEATSTROKE AND HANGOVERS

It can be a blurry line, shimmering in the air, between a really bad hangover and a dangerous case of heatstroke. At a certain point, the division is wholly meaningless, and the symptoms combined become so much more than a sum of their parts—like subatomic fusion.

I realize now that even before this day of lizard tongues, I've experienced similar chain reactions—twice, in fact. Both were south of the border that we're still trying to get to. And both involved a tremendous amount of tequila, in a helluva lot of heat.

One took place in a different lifetime—on a beach in Oaxaca, where I first broke up with the woman for whom I would one day move to an Italian town infested with Ferraris. It was monsoon season, and as far as I can figure, amid the intensity of everything—atmospheric, alcohol-steeped and passion-fueled impossible decisions—my brain and body overheated, self-destructed and basically imploded. Then the woman I'd just broken up with oversaw my dripping carcass from beach to medical station, to hospital, to an expensive private clinic—from which I emerged with a new understanding of relationships, hydration and human fallibility.

The other was just a few years ago, after a good friend of mine got married without a hitch on a beautiful beach outside of Playa de Carmen, and then everyone went home except for me. I stayed an extra day to get to know the town. So then, the day after that, like Elpenor, I woke on a blistering roof overlooking the ocean—only I didn't fall off and break my neck. I got on a bus bound for the airport instead, with a massive sombrero I'd bought for Wasko. It was getting near his birthday, and I'd been feeling drunk and magnanimous. But now, in the morning, locating my seat on the sweltering bus, I felt dank and bilious, combustible and sweaty, saturated and volatile—and then the bus began to move.

One way to fully face the limits of your soul is on a hot, crowded *segunda* bus, surrounded by well-dressed Mexicans so polite that they barely flinch when you retch, again and again—into the vomit-filled sombrero that sits upon your lap.

WHEN LIZARDS DRINK FROM YOUR EYES
(AND YET YER STILL ALIVE)

Sometime later, when the valiant Impending Groom has hitch-hiked back from Tucson with a jerry can of gas, Ted drags me back to the SUV. And by the time we hit highway speed, the air conditioning is working again—like the whole burning world is nothing at all. I can feel the animals inside me slowly reviving, but don't want to ask Chip! to pull over again. So I just dig in, curl up and listen to the voices around me.

It seems there's a whole new problem: apparently, the Impending Groom has waited until now to acquire a marriage license, and the office is about to close. This oversight has come to light through various routes—mostly his brother forgetting to put gas in the SUV—but between grumblings in the front seat and phone calls with those concerned, I can see what's going down: already run over, I am now being thrown under the bus. Either that or there is some other reason to keep muttering my name during crisis-management phone calls. I decide to let it slide, and offer up my hangover to the future of the happy couple.

The new plan is to drive to a tack shop off the highway, where the bride-to-hopefully-be will meet us and take my place for the race to the licensing office. What might become of me is entirely unclear.

When the SUV drives off, I sit for a while by the side of the road, among a hundred bags of dried oats, until finally, out of the Arizona heat waves, a van full of bridesmaids appears.

YOUR WORST HANGOVER EVER

Famous Shakespearean actors have vomited on stage. Boris Yeltsin was found early in the morning outside the White House, in his underwear, trying to hail a cab to get some pizza. And there are several reports—though difficult to verify—of an Indian man who awoke, after a night of heavy drinking, to find he was being digested by a giant anaconda.

But then there are those on a whole other plane: hangovers so heavy they might have changed the fate of the world. In Susan Cheever's recent book *Drinking in America: Our Secret History*, the hours, minutes and milliseconds surrounding the death of JFK are looked at anew: through the blurry, bloodshot, sleep-deprived eyes of the Secret Service. And revealed among grassy knolls, Mafia hit men, slowed-down footage and rifle shots still echoing is this simple probability: that most of the men trained and sworn to protect the President were, in that fateful moment, very much hungover—or possibly still drunk.

The night before in Dallas, many of those in the Secret Service had kept on drinking after last call. They'd gone from the down-town bars to a shady late-night joint known as the Cellar, which served illegal 190-proof moonshine. Six of them stayed until at least 3 a.m., and one until after 5. Their details to protect the leader of the free world and the embodiment of a hopeful future all began before 8 a.m. And before lunch hour was over, Kennedy was dead.

There is no way to know for sure if anyone's hangover that day changed the course of history. But as Chief Justice Earl Warren himself said to the head of the Secret Service, "Don't you think that if a man went to bed reasonably early, and hadn't been drinking the night before, he would be more alert than if he stayed up until three, four, or five o'clock in the morning, going to beatnik joints and doing some drinking along the way?"

In such a light, the slow-motion footage of those terrible moments seems so much horribly slower. The car should be moving forward, swerving, not a suddenly sitting target. The elite should be springing into action. In her book, Cheever notes that those who had been at the Cellar the night before "seemed paralyzed for a few moments." Among them was Clint Hill—whom you've probably seen crawling across the back of the limo. He is reaching for the First Lady, her hands wet with brains and blood.

"It was my fault," Hill would later insist, breaking down in tears during a 1960s TV interview. "If I had just reacted a little bit quicker . . . I'll live with that to my grave." And he wasn't alone—so many of his brothers waking every morning into a changed new world, and the kind of hangover you don't shake off.

As with writers, chefs and private eyes, heavy drinking has long been an integral, even clichéd, part of Secret Service culture, but with heavier repercussions, even when no one gets hurt. Just recently, an agent made headlines—and a hungover trip back home—after being found not standing guard before a hotel door, but instead passed out on the hallway floor.

As Cheever points out, with precise, solemn irony, the creation of the Secret Service, meant primarily to protect the American president, had been approved by Abraham Lincoln on the very day he was assassinated—shot in a theater balcony later that evening while his presidential bodyguard took a break in the bar across the street.

I HAVE SPENT so many years now asking people about their worst-ever hangover. The stories are hardly ever simple. They have to do with terrified chickens, crashed BMWs, jailhouse torture and babyproofing companies; videotapes, martial arts, war crimes and charity canoe races; divorce court, frozen vomit, dissertations and poison ivy. They are devastatingly sad, or horrifically violent, or deeply humiliating, and then some are just plain funny.

One of my favorite hangovers ever I got from a great gal I'll call Jenny. She'd been driving in morning rush hour, trying to get to work after a friend's ill-planned birthday party, stoically steering through waves of nausea. Then two things happened: her bowels started to spasm, and the traffic stopped moving. It was barely a mile to the next turnoff, but a mile is very far when you're traveling an inch a minute and your guts are about to explode. After too many minutes, there was nothing else to do. Jenny looked around—at all those other people, sitting in their other cars—then stared straight ahead and let it happen.

Sometime later, she reached the turnoff, pulled into a gas station, got out of her car with the utmost care and waddled in a sort of bowlegged tightness all the way to the bathroom. After doing whatever she could in the confines of a cubicle, she wrapped her heavy underpants in a shroud of paper towels and put them into the garbage receptacle. Then (and it's partly because of details like this that I've adored her for all these years), she located a pen, wrote "I'm so sorry" on a piece of paper and folded it over a twenty-dollar bill, which she placed on top of the wrapped-up poop.

What I might love most about this story is that Jenny never had to tell it—just as she didn't have to attempt amends with some unknown custodian. But in the hopes of our enjoyment, and perhaps a cautionary warning, she was so kind as to confess. And this was years before I even started asking for such stories.

IN ANY SEARCH for worst-ever hangovers, there are few sources better than ER case reports. And among them, one of the best is "Saturday Night Palsy or Sunday Morning Hangover? A Case Report of Alcohol-Induced Crush Syndrome," about a patient suffering from pain and paralysis in his right arm after drinking a pint of vodka and four pints of lager, then passing out with his arm draped awkwardly over a suitcase. According to the attending doctors' report, this particular method of sleeping off the drink happened to cause a type of rapid, intense muscle decay most commonly associated with "contusions, fractures, or being buried under debris" (or, less commonly perhaps, body-drainage massage), often resulting in renal failure and death. The quick, complex surgery that saved the man's life had been developed for "the treatment of victims of natural disaster or war."

Then there's "Retained Knife Blade: An Unusual Cause for Headache Following Massive Alcohol Intake," in which a twenty-two-year-old man was brought to the ER by his relatives, who referenced a family binge of rum and beer but were otherwise unhelpful. He stunk of booze and was uncommunicative, except for a sort of growl. After treating him with intravenous hydration and multivitamins, the medics monitored his sleep "in expectation of sobriety." Finally waking eight hours later, the man complained of a headache—a really, really bad one. Following extensive neurological examination and heavy intravenous pain-killers, including morphine, the pain persisted, unexplained. So they scanned his skull and found what's bound to make a bad hangover your worst one ever:

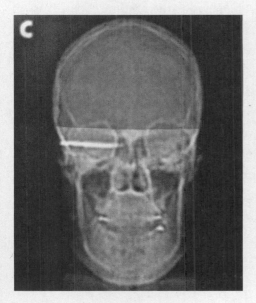

If you're not sure what you're looking at, you might refer to the case report's title and then that long white line just behind the eye.

WHEN LIZARDS DRINK FROM YOUR EYES
(AND THE BRIDESMAIDS PICK YOU UP)

Of all the horrible and humiliating situations in which to find yourself unmanageably ill, soaked in your own stinking filth in a van full of pretty bridesmaids is a unique take on hell.

And as if to witness what they wrought, the hiccups choose here and now to make their insipid return. This, of course, the bridesmaids find hilarious—until I start to gasp between hiccups for the van to stop. Then, as I try to step out onto the side of the road, it's like my ankles aren't even there anymore; I tumble into a ditch and lie at the bottom, heaving out scrapings of bile.

When at last I make it back to the van, there are no more giggles—just a silent, amazed revulsion. The rest of the drive is quiet, until finally—six and a half hours after leaving the hotel in Tucson, just sixty miles north of here—I arrive at the resort.

The mother of my boy-to-be, bless her heart, decides not to kill me. Probably something to do with the Hippocratic oath— she is, after all, a doctor. On the bed, in our room, she takes my temperature and audibly gasps at the reading. She turns on the water for the bath, unwraps my ankles, takes off my clothes, listens to my heartbeat and shines a light in my eyes as I tell her about the lizards. Then she helps me into the bathroom and into the tub—which is full of cold water. My body sizzles as the breath rushes out of me.

After wrapping me in wet towels on the bed, she starts getting ready for the rehearsal dinner. I tell her she looks nice. She rolls her eyes, pats our baby in her belly and hands me the remote.

"That *Hangover* movie is on pay-per-view," she says. "You might like that."

🍾🍾🍾

WHAT *THE HANGOVER* DID

To briefly recap, *The Hangover* is a 2009 film about an ill-fated bachelor party and the groom's friends from out of town—including the wacko brother of the bride, a well-groomed professional and a guy so nondescript he'd make a pretty good spy. They raise a glass to the groom at their hotel, then the rest is madness and mayhem.

It involves erotic dancers, newborn babies, sports cars and a Bengal tiger; hotel managers, cops with handcuffs and incriminating videos; a fetal lunatic in nothing but his socks, a breakdown in the desert, heatstroke, a rooftop . . . and don't forget that tiger.

It is a brilliant, and arguably prescient, film, but there was still no sober, predictable reason for the mind-blowing degree of its success. It's not like the subject broke new ground. The fallout from prenuptial debauchery is, after all, terrain so well trampled by Hollywood—from *The Philadelphia Story* to *Bachelor Party*, *Very Bad Things* to *Wedding Crashers*—that it is practically a genre unto itself. But nothing, including the crummy sequels, can diminish the legacy of Todd Phillips's original masterpiece.

Not only did *The Hangover* break every box-office record for an R-rated comedy, slingshot Bradley Cooper, Zach Galifianakis and Ed Helms onto the A-list, spark a Hollywood renaissance by reminding production companies that great comedy requires high risk, and even provide Mike Tyson a much-needed bit of levity, but it also changed the modern culture of hangovers.

Dr. Burke, of Hangover Heaven in Las Vegas, suggested that our present preoccupation with the condition was cue to three concurrent semi-revelations: state-sponsored statistics on the cost of hangovers to the economy, a study by Dr. Jeremy Wiese of New Orleans suggesting the remedial efficacy of prickly pear and—of course—*The Hangover*.

And it's possible that those first two bits of news might have gone practically unnoticed if not for the overwhelming, undeniable, somewhat surprising effect of *The Hangover* on our vast collective consciousness. More statistical studies on hangovers were started within a year of the film's release than in the previous century, and a record number of hangover products hit the market. Along with Chaser, made by Manoj Bhargava, and NoHo, created by Dr. Wiese, one of the most successful sellers, at least for a while, was Hangover Joe's Recovery Shot—the label boasting "Officially Licensed by the Movie *The Hangover*."

Another thing the movie did, in a psychologically roundabout way, was normalize the morning after. Not only did it set a bar (albeit fictional) so impossibly high (or maybe low) that most human hangovers seemed comparatively innocuous, but the now-aging genre of hangover stories had a new, all-encompassing reference point.

Almost a decade later, a basic internet search for anything to do with hangovers is still an exercise in filtering out an ocean of marketing, buzz lists and blogs about the *Hangover* trilogy. And then what you're left with is a swamp of products and personal confessions—most of which still reference the movie as a way of relativizing the experience—even more so when it comes to the endless trove of wedding-day hangover reportage. It is safe to say that at no time in history has there been a more comprehensive record of people throwing up in church.

It's also worth mentioning that the limit-pushing, far-reaching success of *The Hangover* depends, in part, on a happy ending and a wedding saved. But in life, of course, it doesn't always go that way. Even among so many daily miseries, one of the most depressing stories ever run in the British tabloid the *Sun* is surely that of Siobhan Watson's wedding day.

The day began with Siobhan waking in a strange budget hotel room at 12:30 p.m.—just a half hour before she was supposed to get married. Absorbing in a flash this awful information, she lurched to the bathroom, vomited, then promptly passed out for another hour. And in those sixty little minutes—her rosy cheek pressed against the bathroom tiles in a baby-like puddle of drool—Siobhan's entire life was changing.

One can only imagine how it must have felt: to wake once more, to stagger and stand, then read and comprehend, the numbers on that bedside digital clock . . .

For years, Siobhan's parents had worked and saved so that she could have a dreamlike wedding—the culmination of every happy inspiration, any vision of the future she'd ever conjured up. For the past few months, she'd had trouble falling asleep—not from anxiety, but excitement; she imagined it all in vivid detail, even the script of her father's speech, right down to the proud, silly jokes he'd make. The future seemed limitless, wonderful, so very close—and she'd giggled and cried in bed, just thinking about today.

And now it was here. And now it was gone.

She wasn't even a drinker, didn't even like the stuff. But far more than that, she hated disappointing people—and so had finally given in to her friends when they begged to take her out for bachelorette drinks the night before. And then the drinks hadn't tasted too strong . . . but by the third one, Siobhan was drunk, for the first time ever. And the fourth one blacked her out. No one would ever know what happened after that—how she ended up here, passed out in this room, instead of at her wedding.

After a while, Siobhan left the hotel and started to walk. The shock kept her legs moving—without rest, water or a sense of direction—for five full hours while family, friends and local police searched the city streets. And though she finally did show up, just as the sun was setting, there would be no happy ending.

Her fiancé broke up with her, and then it took her parents a formidable while to let her move back home—a wedding and future so beautifully conjured up, then simply made to vanish.

"I can't believe I ruined it," Siobhan told the *Sun*, "by having one too many piña coladas."

WHEN LIZARDS DRINK FROM YOUR EYES (AND STILL YOU RETURN TO VEGAS)

After watching *The Hangover* like some kind of Pavlovian aversion therapy—shaking under wet towels, laughing and coughing up drops of leftover bile—I watch it again when my gal gets back from the rehearsal dinner.

"You're really going to like this," I tell her.

"You really should be dead," she says. And then we snuggle up.

She's right, of course. And it's this that can make the worst of hangovers dangerously invigorating—like coming out of a fight, or an unspeakable loss at the poker table. It feels like you should be dead, but somehow you're still alive. And for a little while, in the painful afterglow, anything seems possible.

After the wedding, we'll drive together down Route 66, back to Las Vegas, to catch our flight home. I don't know it now, my hand on her belly as my darling watches *The Hangover*, but when next I return to these deserts, it'll be as a single dad—with a buried friend, a new book to write, a new girlfriend to lose and a spot reserved in Hangover Heaven.

THE HANGOVER WRITER

For art to exist," said Nietzsche, "a certain physiological precondition is indispensable: *intoxication*." Or as Horace put it a couple of thousand years earlier, "The dulcet Muses have usually smelt of drink first thing in the morning."

It's never been a secret—and is by now more a hackneyed, tired truism—so much so that Sir Kingsley Amis himself, in the greatest-ever essay on hangovers, deigned only to dismiss it, suggesting that writers remain drunk not because of artistic temperament or creative process; instead, it is "simply that they can afford to, because they can normally take a large part of the day off to deal with the ravages."

But while that might hold water, or some kind of Veuve, for the Lord Byrons, Lady Woolfs and Sir Amises of the realm, it's a lot different for the Bukowskis, Highsmiths and Carvers in the street. What one "can afford" is also a question of survival. It's like a philosophical photo negative of sober Pliny's conceit that drunkards lose "not only the former, but also the coming day." A drunkard who's also a writer might sacrifice safety, security and even the dawning sun to keep diving for the depths of the night—to salvage them for the page.

"The writer walks out of his workroom in a daze," wrote Roald Dahl. "He wants a drink. He needs it. . . . He does it to give himself faith hope and courage. A person is a fool to become a writer. His only compensation is absolute freedom. He has no master except his own soul and that I am sure is why he does it."

In psychological terms, that "faith hope and courage" might be termed "disinhibition"—or what William James called the "great exciter of the 'Yes' function." If anyone needs a function like that, it is the poor lout trying to create something meaningful while every next word sets off a chorus of "No!" One of the many things that booze can do, with all its yes-ness, is negate the need for perfection while also making whatever you're trying to do feel somehow meaningful. So, in such a state, you may actually manage to write, at least for a little while. And of course, as Dudley Moore said in Arthur, "Everyone who drinks is not a poet. Some of us drink because we're not poets."

At the other end of the spectrum from Arthur was Edgar Allan Poe, a soul so obsessed, skilled, haunted and unable to handle his drink that it both killed and immortalized him. According to all available sources, Poe's connection to booze was as intense and disturbing as his work: a dark, innate chemistry that made him at once immediately ill and entirely unable to stop. He didn't keep a diary, never wrote about such things directly, and yet the everlasting world that he rendered—of impending doom, prolific anxiety and the visceral horror of being buried alive—seemed to come from a maelstrom of unending hangover.

If any writer understood this, it was Malcolm Lowry, the brilliant, besotted author of Under the Volcano, who once explained a passage in which Poe detailed the desire to be "led captive by Barbarian hordes, on some Arctic island of wilderness and snow, in a land desolate and unknown" as "Poe's admirable description of hangover."

For the ultimate *actual* description of hangover, we need only return to Amis. At the very beginning, in his debut novel—long before being knighted by the Queen—he woke up Lucky Jim Dixon.

> *Dixon was alive again. Consciousness was upon him before he could get out of the way; not for him the slow, gracious wandering from the halls of sleep, but a summary, forcible ejection. He lay sprawled, too wicked to move, spewed up like a broken spider-crab on the tarry shingle of the morning. The light did him harm, but not as much as looking at things did; he resolved, having done it once, never to move his eyeballs again. A dusty thudding in his head made the scene before him beat like a pulse. His mouth had been used as a latrine by some small creature of the night, and then as its mausoleum. During the night, too, he'd somehow been on a cross-country run and then been expertly beaten up by secret police. He felt bad.*

Sir Kingsley, no doubt, felt this particular kind of *bad* countless times over the next forty years, the writing of twenty more novels and, eventually, that hangover essay. In the final chapters of even the most loving and authorized biographies, we see the great man brought down—staggering, bloated, babbling. As Zachary Leader put it, "The booze got him in the end, and robbed him of his wit and charm as well as his health."

For whatever reason, the only authors of note to write entire books (albeit very slim ones) on the subject of hangovers since Amis's essay were also rather gluttonous Englishmen.

Clement Freud—broadcaster, theater producer, member of British Parliament, nephew to Sigmund and author of

Hangovers—was a famously sophisticated upper-class bon vivant. And Keith Floyd, who wrote *Floyd on Hangovers*, was a garrulous, gambling man of the people—restaurateur, raconteur, foodie, fabulist and a TV celebrity chef before that was even a thing.

Like Amis, both men lived adventurous, eclectic, somewhat Dionysian lives. But while Freud's passing at the age of eighty-four was noted with warmth and respect, Floyd went down, twenty years younger, with the sort of mishap, messiness and irony of a true hangover writer.

"Keith Floyd, the chain-smoking, hard-drinking television chef, has died from a heart attack only hours after a lunch to celebrate being given the all clear from bowel cancer," reported the *Telegraph*, which then listed every item of Floyd's final, boozy, celebratory brunch, along with each cost—not unlike the list of his failed restaurants and marriages so thoughtfully provided in the accompanying obituary.

AMONG COUNTLESS HUNGOVER writers, so very few have written actual hangover books. And I'm not sure why I chose to join them. It's a question that has started to haunt me.

A quarter century back, when I first started drinking and writing and indulging in all sorts of rambling, I stupidly, typically, saw myself as a modern incarnation of Jack Kerouac—like all those maniacs throughout history imagined themselves as Dionysus. But just as this whole book here can be summed up with the question "What could possibly go wrong?" the answer is even more pat, and patently unhelpful: "Be careful what you wish for."

Still, after all these years, young Kerouac disciples seem oblivious to how fleeting was his time intoxicated and enraptured on the road, compared to coming off it. When *On the Road*

was finally published, he was already a man who dealt fully in
regret and lived in a state of perpetual hangover.

A few short years before drinking himself to death—his belly
full of Aspirin floating on a sea of whisky—Kerouac wrote *Big
Sur*, at the age that I am now. And this is how it begins:

> *The church is blowing a sad windblown "Kathleen" on
> the bells in the skid row slums as I wake up all woe-
> begone and goopy . . . drunk, sick, disgusted, frightened,
> in fact terrified by that sad song across the roofs mingling
> with the lachrymose cries of a Salvation Army meeting
> on the corner below . . . and worse than that the sound
> of the old drunks throwing up in the rooms next to mine,
> the creak of hall steps, the moans everywhere—including
> the moan that had awakened me, my own moan in the
> lumpy bed, a moan caused by a big roaring Whoo Whoo
> in my head that had shot me out of my pillow like a ghost.*

And that is the scariest part of finally becoming a hangover
writer: it's not the tactile—the goopy, moaning sickness. That is
but a diversion from the big roaring Whoo Whoo, for which
there'll never be a cure. It's what shoots you out, no matter
what—a ghost into the day.

Part Eleven

AFTER THE FLOOD

IN WHICH OUR HUNGOVER WRITER WASHES UP IN
NEW ORLEANS, CONFERS WITH THE UNDEAD, VISITS
THE LIMITS OF THE MORNING AFTER—AND FINALLY
COUGHS UP A CURE. APPEARANCES BY VAMPYRE
MARITA JAEGER, DR. MIGNONNE MARY
AND VOODOO PRIESTESS MAMBO MARIE.

I tried to drown my sorrows,
but the bastards learned to swim.
—FRIDA KAHLO

I AM SIPPING A VAMPIRE'S KISS, SITTING ON THE UPPER outdoor balcony of Muriel's Jackson Square—famously haunted by the self-murdered man who built all this: the parlors, drawing rooms, bedrooms, wine cellars, dining rooms, gardens, game rooms and bars that were going to be his home, before he gambled it all away in the breadth of a single night.

In the busy square below, there is a woman with a fiddle, a man on a ragtime piano. Their fingers are flying, wild hair describing halos—and they are both singing some operatic, leaping reel. Just a few feet away, in another circle in the crowds, a man-as-self-made-robot transforms impeccably, impossibly, into a four-wheel jeep—rolling in time to the music as it rises toward me. I can feel last night's absinthe and Toxic Baby like ghosts in my blood, but also sunshine on my face.

This whole city, of course, is haunted: by those who were brought as slaves, went up in the fires, down in the rising floods. And it is the haunting that is everything: meaningful and monstrous, beautiful and brutalized—restless souls let back into the party, brought back to life through the music and booze of the living.

So much that has changed the course of hangovers, and humanity's relationship to the bottle, has docked in the port of New

Orleans, itself like a latter-day Amsterdam: cognac and cholera, absinthe and opium, science and slavery. This city is a living history of how and why the world has gotten drunk: all the rituals, celebrations, diseases and detritus from across the oceans; two hundred years of disparate, desperate souls from every corner of the globe, roaming the bayou, dropping seeds in the swamp, where the most potent, wondrous things have then grown. They are what brings one here—to these gardens of good and evil, and houses of the rising sun.

TO SUMMARIZE THE last long while in one single, nearly balanced line: I found what I was looking for, and lost almost everything else.

Among the things that still remain are this book and the remedy. I could, of course, just type out the recipe now—and finally be done with it. If you've stuck with me all this way, there's no doubt that you deserve it. But there are still a few things I have to do first—for the completion of this story, the history of hangovers and comprehension of the cure. Also, I think the cure might be cursed . . .

"Your table is ready, sir," says a voice. And I am led inside, to a setting of turtle soup and strawberry salad, shrimp and grits and goat cheese crepes, Dixieland jazz and flutes of chilled mimosa. It all works as one might hope: steadying the system and quieting the ghosts.

I got drunk last night in Pirates Alley, for three dollars a shot. "What's in the Toxic Baby???" read the label on the bottle, the bottle on the bar. "A proprietary blend of the most cost-effective ingredients . . . For three bucks you don't need to know." This place is all about mysterious elixirs.

After brunch, I walk down the sun-drenched street to the dark door of Boutique du Vampyre. It carries all sorts of kitsch for the undead, and shop owner Marita Jaeger's teeth are filed into perfect fangs. I ask her what vampires do for a hangover.

"Hmm," she says with a half smile. "Like if you've drunk too much from the veins of a drunkard?"

"Sure," I say, and she points me to the "Energy Blood"—a thick red goo packaged in medical transfusion bags. It is full of iron, electrolytes and caffeine—and apparently it tastes like tropical punch.

Then, on the next shelf over, I spot something even more alluring—a row of glass bottles that read:

DOC. CHESTER GOODE'S UNIVERSAL ELIXIR,
FOR NOCTURNAL CREATURES.

On the label beneath is a picture I recognize: William Hogarth's fantastical rendering of a hellish Gin Lane—and below that, an explanation: "For all that ails you! For Medicinal Purposes only!"

But then, to my undying disappointment, I find that the bottles are empty.

"The idea," Ms. Jaeger tells me, the tip of her tongue snaking between fangs, "is to find your own concoction and fill it up with that."

A UNIVERSAL ELIXIR, FOR NOCTURNAL CREATURES

This is, of course, what I set out in search of, so very long ago. And in some ways I'm still looking. It's not that I haven't found a cure that works for me; I have. It's just that the efficacy of my elixir for my nocturnal habits depends entirely on it being taken nocturnally. To save myself from the morning after does require a moment of prevention the night before—just a modicum of forethought.

Remedy in hand, I've come to talk with Dr. Mignonne Mary, founder of the New Orleans Remedy Room, for her opinion on what I've got. Yet still there's that voice from my lizard brain: *But what about the creatures who've forgotten to take it—too drunk and stupid, passed out in the forests of the night? What hope is there, then?*

So I am also here, in New Orleans, for one last dig—for when a nocturnal creature like myself neglects to take the elixir. Muriel's Sunday brunch might do the trick—at least better than an empty bottle. But I've still got some places to go, and a few last things to nail down about the morning after, before divulging how I stop it from ever starting.

AFTER THE FLOOD (HAIRS OF DOGS AND TAILS OF COCKS)

I have, over the course of the past few years, been given so many hangover remedies: from kimchi to kombucha, Advil to acupuncture, haggis to heroin. But everybody is different, and so is every hangover. If the turtle soup works today, it might not work tomorrow.

And even if something is effective, it's never a magic morning bullet but rather a matter of speeding up the healing process: reducing inflammation, settling stomach acids and making sure the body can retain fluids again. And since a hangover rarely lasts, at least acutely, longer than twenty-four hours, the parsing of that time into half-lives becomes like splitting hairs of the dog—which is still the most popular way to burn that beastly time.

Although often problematic, the hair of the dog is borne out by experience and little bits of science. And there's nowhere it's been tested more than here in New Orleans, the birthplace of the cocktail—or at least where it was fostered and brought to full fruition. This was due in part to all the rum, sweat and sweetness that came out of the sugar plantations—but, even more so, that good Louisiana ice. After all, the tail of the cock was first just a fancy hair of the dog—and who wants a warm one of those?

For most of human history, the only way to get ice in your drink was to chip it off a glacier or frozen lake—or have someone else do it, send it to you on ships and trains, put it in a room with sawdust, and then mix all the drinks you could before it melted. And because of the way the world and weather worked, the ice harvest only lasted about six weeks a year. The rest of the time, you were sipping something warm—a quotidian way to drink at night, and just awful in the morning.

In 1840, right here in New Orleans, the first-ever commercial icehouse opened up. And pretty soon, every saloon in the French Quarter had a whole cache of ice shakers behind the bar, shaking up Ramos gin fizzes and Corpse Revivers, French 75s and bourbon sours. These were known as eye-openers, pick-me-ups and hairs of the dog. Now they're just called cocktails.

And do they work? You bet they do. They work subjectively and poetically, to paraphrase Barbara Holland, by curbing the sickening down-swoop of alcohol leaving the system. And they work scientifically and objectively in the way that Adam Rogers explained—hindering the by-production of methanol by taking in more ethanol.

But might the hair of the dog also work philosophically—or, more precisely, Eleatically? If you don't know what that means, I'm sure it doesn't matter—as long as Josh Parsons does. During his research fellowship with the Arch Centre at the University of St. Andrews, Parsons wrote "The Eleatic Hangover Cure," a possibly seminal paper published in 2006, in which he sets out to prove—purely theoretically—that "there *is* a way to completely cure a hangover using a hair of the dog."

Starting with the supposition of there being "a simple and direct relationship between the amount of alcohol consumed and the duration of its effects," in which each drink will induce an hour of intoxication followed eventually by an hour of hangover, so he instructs: "Start drinking at a time when you are not drunk, and have no hangover." The idea, then, is to drink half your glass, wait for half an hour—"until you are just about to get a hangover"—then drink only a quarter of it, wait for fifteen minutes, and so on down to the tiniest sips divided supposedly by seconds, until a last invisible drop when the hour is up and your glass is finished, so that "every incipient hangover you might have had was cured by a further drink."

There are, of course, many ways to counter Parsons's polemic. But in his paper he confronts them so good-naturedly, in a series of specific rhetorical questions, that by the end, one might admit it that *could* be true, at least philosophically—despite how extremely annoying it would be to drink your drinks like that.

THE GREEN FAIRY AND THE BITTER TRUTH

No history of hangovers could be complete without a discussion of the two most vilified of drinks: absinthe and Jägermeister. And ironically, the empty elixir bottle from Ms. Vampyre Jaeger might just be filled by understanding them.

In the New Orleans Museum of Pharmacy there is a big old bottle of bitterroot. According to the accompanying text, "An Egyptian papyrus dating to approximately 1300 BCE mentions its use as an ingredient for soothing stomach ailments." I wonder if Dr. Nutton knows in which landfill it was found; it might pair nicely with those drrrrunken headache-relieving leaves of Alexandrian chamaedaphne. Also, according to the museum, "It is believed that bitterroot was an original ingredient in Absinthe."

Traditionally made from all kinds of plant-based ingredients, including anise, fennel, hyssop, lemon balm, bitterroot, licorice, angelica, marjoram and wormwood, there is no tonic in modern history more mythologized and maligned than absinthe. The green glow of chlorophyll set free from some of those herbs looked like magic in a bottle, and some people who drank it were said to get more than just drunk.

Known as *la Fée verte* (the green fairy), it was both muse for and subject of a whole whack of masterful Bohemians: Byron, Poe, van Gogh, Toulouse-Lautrec—of whom Harry Mount wrote in the *Telegraph*: "He painted like a hangover feels—like the worst hangover of all, the absinthe hangover, brilliantly designed for nausea and remorseful over-examination of the night before. In his pictures, the skin of the dancers and patrons is yellow, absinthe-green and a ghostly white." It became the drink of Paris, and then New Orleans.

As the absinthe exhibit in the Museum of Pharmacy puts it: "People hospitalized from consumption of absinthe were observed to have seizures, chest effusion, reddish urine, kidney congestion, visual and auditory hallucinations and a high suicide rate." It was the wormwood that took the blame. Scientists pointed out that it contained thujone—a chemical compound that in large amounts could mess with your head, nervous system and kidneys. And non-scientists pointed out that it was called *wormwood*—which sounded creepy. So absinthe and anything else containing wormwood was outlawed across the ocean, for five years shy of a century.

But during all that time, New Orleans held a torch, cleaned the fountains, kept the faith, as it does. And now, of course, we know that, like bitterroot, most of the ingredients, including wormwood, can be greatly beneficial. And in fact, the cause of "severe absinthism" was most likely the same as it was for the wretched toxic madness during Prohibition: chemical contaminants added to the cheaper stuff to make colors and flavors more appealing—that, and what sometimes happens if you down enough 140-proof booze.

The truth is, if you are looking for something to put in a morning-after tonic, it's probably bitter herbs. But during, and possibly due to, the hundred-year absence of absinthe, it seems we lost an understanding of their benefits, and how they organically combine with alcohol.

For centuries, bitter herbal distillations were the ultimate connector—of medicine to intoxicant, earth to water, root to the expanse and limits of humanity's ability to deal with natural, magical potions. And after absinthe, no green glass bottle so illuminates the modern breach in our understanding as that of Jägermeister.

As our Vampyre Jaeger, just down the street, would surely explain, *Jäger* means "hunter," and *Meister* means "master." Jägermeister was conceived for local woodsmen: to heal the body, warm the bones and soothe the soul after the killing of a beast, and then calm the gut before the feast. Now sold in 109 countries, it has become one of the most recognizable, yet misunderstood spirits in the world.

Of the fifty-six herbal, fruit, root, spice and floral ingredients, only twenty-five have ever been divulged to more than a handful of people. At the same time, Jäger drinkers have become synonymous with a sort of sensory thuggishness and bro-culture idiocy. Much of this, of course, is a result of mixing it with Red Bull into mindlessly tragic Jägerbombs and bingeing on something that was meant to be taken as a carefully constructed elixir.

So, when complex, crafted bitter emulsions like absinthe and Jägermeister are blamed for infinite, unspeakable hangovers, it is probably a matter of taking what could be a potent remedy not only for granted, but taking also way too much of it.

AFTER THE FLOOD (AND WAY TOO MUCH OF EVERYTH NG)

A lot has happened in the last long while, and I've lost many things: loved ones, friends, family, jobs, homes, the simple sense of self and a firm understanding of the passage of time. I've also done a lot: traveled to Berlin and Mexico and into the past—through detox, rehab and various kinds of therapy. I've subjected myself to clinical trials and created ones of my own. I have, in fact, tested, tweaked and tried my own concoctions so many times, in so many ways, that I can now say with confidence that, yes, I have found a cure for my hangovers—or, more accurately, an antidote, or perhaps a prophylactic. But essentially, it is a cure.

I have a mix of ingredients and a method that, used properly and at the right time, appears to prevent all acute symptoms of hangover in me—nausea, gastric distress, vomiting, headaches and muscle pain—leaving only the less tangible ones, like exhaustion and lethargy.

The truth is, I never really expected to find it. Not really. And then, when I did, it coincided so directly with my life falling apart that I didn't know what to do with it. I became like a man scared of himself, yet undaunted by the morning. And that is very dangerous.

But isn't that often the way with quests? The hardest part is knowing what to do should you somehow reach the end.

So afraid was I of what I found that, for a while, I gave up drinking. But of course, it was still out there: my cure, just waiting for me to return.

To quote Dr. Jekyll:

> *I was true to my determination; for two months, I led a life of such severity as I had never before attained to, and enjoyed the compensations of an approving conscience. But time began at last to obliterate the freshness*

of my alarm . . . and at last, in an hour of moral weak-
ness, I once again compounded and swallowed the trans-
forming draughts. . . . My devil had been long caged, he
came out roaring. I was conscious, even when I took the
draught, of a more unbridled, a more furious propensity
to ill.

Like the fallible doctor once more overcome, the fictional scientist at the mercy of his own study, I watched myself submit
again to that deep-down thirst for experiment, the possibility of
transformation.

I drank my drinks and took my pills and wreaked all kinds
of havoc. I stormed across the globe, lost my passport and fell
into canals. I opened a bar in an underground wine cellar and
named it the Lowdown. I booked great bands and wondrous
DJs, slung drinks, invented cocktails and named them after Tom
Waits songs. I drank with everyone and never got a hangover. I
missed my girlfriend, missed so much, and things just kept on
falling, into smaller and smaller pieces, little bits of loss, until I
couldn't even write them down.

THE SNAKE OIL PARADOX

"Snake oil salesman" has, of course, become shorthand for all
sorts of grifters and fraudsters—and I've heard it used for pretty
much everyone who's ever marketed a hangover cure. But what
would it mean if snake oil actually worked—which it almost certainly does?

Used for thousands of years, the oil of the Chinese water
snake was introduced to the New World when indentured
Chinese railway workers—who took it to soothe the pain and
inflammation caused by brutal manual labor—shared the stuff

with their North American coworkers. It worked so well that the makers of patent medicines took notice.

The bad rep came with hucksters like Clark Stanley, the Rattlesnake King—a cowboy who spun snakes in the air like lassos and sold Stanley's Snake Oil from town to town until the Food and Drug Administration seized his wares in 1917 and found that it contained mineral oil, beef fat, red pepper and turpentine, but not a lick of snake oil. And so the snake oil salesman became synonymous with opportunistic bullshit not because he sold snake oil, but because he actually didn't.

THROUGHOUT MUCH OF modern history, at least in the West, the traveling medicine show was a stunningly popular melding of circus entertainment, scientific experiment, freak show, medical lecture and infomercial. The products, known as patent medicines, were for the most part bitter, usually boozy elixirs full of vitamins, minerals and sometimes cocaine, opium and even bovine blood.

With names like Moxie Nerve Food, Dr. Kilmer's Swamp Root and Dromgoole's Bitters, they were sold as panaceas for whatever might possibly ail you. But they were most notably effective in treating "the Irish flu." In fact, the time of the traveling medicine show was probably the last heyday of hangover remedies. And it ended right here in New Orleans, with Dudley J. LeBlanc.

Before becoming the Cajun impresario behind Hadacol vitamin tonic, LeBlanc was a World War I soldier, a traveling shoe salesman, the owner of a pants-pressing service and later a burial insurance company, and also sold tobacco. Then he went into politics, became a senator and set his sights even higher. Between his many bids for the Louisiana governorship, LeBlanc made patent medicines. And the campaign trail was the perfect place to hock his wares.

He started out with Happy Day Headache Powder, a folksy name that seemed to allude to a better morning but didn't really sell—and soon fell prey, along with others like it, to the new FDA regulations designed to put an end to patent medicines. They required that all ingredients be listed on bottle labels, and for licensed physicians to be present at even the most ragamuffinish tonic-shilling sideshow. So for his new product, Hadacol—a much more scientific-sounding name—LeBlanc didn't just list the ingredients, he promoted them, and put a slew of physicians on his seemingly endless payroll.

LeBlanc's traveling medicine shows were epic: both a familiar, folksy throwback and a daring precursor to the star-studded, product-sponsored festivals of today. Lucille Ball, Bob Hope, Hank Williams and dozens of other A-list performers toured the country in the Hadacol Caravan. It was the best party, wherever it went, and all it cost to be there was to buy a bottle of Hadacol—which tasted like swamp water but was sure to give you some pep.

The Hadacol blues and jazz barbecues lasted into the night. By the late 1940s, you could get Hadacol six-gun holsters, shot glasses and a *Captain Hadacol* comic book. And there were literally dozens of honky-tonk, blues and bluegrass songs about drinking the stuff. When Groucho Marx, on his TV quiz show *You Bet Your Life*, asked what Hadacol was good for, Dudley J. LeBlanc quipped, "About five million dollars for me last year."

But in 1951, the American Medical Association issued this statement: "It is hoped that no doctor of medicine will be uncritical enough to join in the promotion of Hadacol as an ethical preparation. It is difficult to imagine how one could do himself or his profession greater harm."

Shortly thereafter, LeBlanc announced that he was selling the Hadacol empire to the Tobey Maltz Foundation for $8 million but staying on as sales manager for $100,000, and that he was making one last bid for the Louisiana governorship.

But as the Museum of Pharmacy puts it:

It soon became apparent to the purchasers that the accounting for the Hadacol business was less than accurate: profits had been greatly inflated and preciously hidden debts resurfaced. LeBlanc was also faced with a complaint from the Federal Trade Commission that Hadacol advertising had been misleading and the Bureau of Internal Revenue charged that he owed $600,000 in taxes. The resulting scandals ended any hopes that LeBlanc had for being elected governor. Several years later LeBlanc invented a tonic called Kary-On to try and revive his fortunes, but it never became popular.

When LeBlanc's Kary-On was seized by the FDA, they found it contained almost exactly what Hadacol had, but without all the booze—that is, a bunch of B vitamins, including niacin, and minerals, including magnesium—pretty much the bedrock on which so many hangover products to follow would also rise and fall. Even the name, Kary-On—cutesy, misspelled and multiple-meaninged—was like a template for all the hangover product trademarks to come.

And the ghosts of LeBlanc's endeavors—with their optimistic hubris, questionable science, boozy vitamins and populist politics sucked down into the snake-oil swamp against the backdrop of New Orleans—now leave a thing that hovers in the air imperceptibly, pervasively, possibly ancient, but stronger than ever, and tastes a lot like curse. And the curse is this: Nobody shall believe.

They'll believe that a supreme being exists. Or that many do. Or none.

They'll believe we've rocketed our way to the moon. Or else have faked it all.

They'll believe that positive thoughts can make good things happen.

They'll believe that ours is merely one in a limitless web of universes.

They'll believe in love.

They'll believe in antimatter—and that if we create enough, we can travel to other galaxies.

They'll believe in capitalism, communism, nihilism, heroism, fascism, artistry, poetry, diversity, forensics and eugenics.

But nobody shall believe in a cure for the common hangover.

AFTER THE FLOOD (AND INTO THE REMEDY ROOM)

"He was always way ahead of his time," says Dr. Mignonne Mary, adjusting the tube taped to my arm, then sitting down across from me, but close. Everything she does seems close, intimate. In French, her name means "cute," which is a drastic understatement. Her voice, all Louisiana peppered, is also low and humming. She is talking about her father.

"My whole life, I've been watching people as they wake up and finally hear the things that he was saying. Saying them all this time."

Dr. Charles C. Mary Jr., the head of a New Orleans hospital by the time he was thirty-two, developed ideas and implemented treatments—particularly pertaining to magnesium—long before they became recognized by standard medical protocols. He was a groundbreaking, often controversial pioneer in many fields (especially regarding the effects of alcohol).

Dr. Mignonne Mary comes to her career, and her interest in all of this, honestly and intimately. Every day, she is using the research, discoveries and hard-learned lessons of her dad—to help people feel better, to make a living and to keep on experimenting.

What is dripping into me now is not unlike what I got from Dr. Burke, all the way back in Hangover Heaven: a Myers' cocktail but with all sorts of added Mary-legacy magnesium. I think about what Sydney Smith wrote, hundreds of years ago now, regretting to himself the catalog of his intake from the night before: "After all this, you talk of the mind and the evils of life? These kind of cases do not need meditation, but magnesia."

"Mag is magic," says Mary. "It is a vasodilator and neuro-relaxant. It is anti-anxiety, opens blood vessels and relaxes muscle walls so fast it can stop asthma. It can stop a pregnant woman from contracting. It's powerful stuff and, taken properly, can sort you out just fine."

It is impossible, of course, on pragmatic levels, to compare the treatment I'm getting now to the one in Hangover Heaven. Not only did I mess that one up—by being drunk instead of hungover, then trying it again after three days of incomparable g-force lunacy, only to find that maybe it worked, though my brain no longer did—but that was years ago now, when my liver was leaner, my body fitter, my mind less troubled and hangovers seemed like a whimsical thing. And also, it was Vegas. This is New Orleans, where hangovers can feel as natural as sunlight and the rising tide. But then again, both those things can also do a lot of damage.

After an hour or so in the Remedy Room, taking the drip, sipping tea and listening to the recorded music of flutes. I feel better than when I came in.

For what it's worth, I'd have to say that IV treatments—be they in Hangover Heaven, the Remedy Room or one of the similar spots now springing up all over the place—are generally quite effective for me. And the truth is, I've tried many of them now, including some administered by off-duty doctors in their living rooms.

Except perhaps for a massive dose of adrenaline, I think this may be the most effective way to curb my hangover once it's already started. But unless you are a doctor and/or have all the required equipment, ingredients and expertise at your disposal, you've got to get to a place like this, or have them come to you—simpler, at least, than jumping off the Stratosphere.

But of course, none of this should be necessary—which is what I really want to talk about with Dr. Mignonne Mary.

"I have something," I say, in my own close way. "And I'm hoping to get your opinion."

"Okay, then," she says, smiling. "Why don't you give me the lowdown?"

THE LOWDOWN

What follows is the best home cure I've devised so far for my own use; however, I am not a doctor, so I cannot recommend that you try it without first obtaining the advice of a qualified medical professional. I can't even claim that this remedy is new. All of these ingredients have been around a long time, and many have been used together for this very purpose. But having tested my concoction against every similar product or method I could find, I've concluded—at least for now—that no other remedy is more consistently reliable for me.

I believe the efficacy is due to the following: (1) the high doses of NAC and anti-inflammatory; (2) the method by which they're ingested—specifically, swallowed in pill form, and at the right time; and (3) the belief, based on informed opinion, that it can work.

But I am sure this concoction can still be improved upon. That, after all, is why I'm here, talking to Dr. Mary.

For now, we'll go back to what I've got and the method by which I take it. First, I acquire the following ingredients—all of which are easily accessible, fairly inexpensive and "natural," insofar as they also occur somewhere in nature:

VITAMINS B_1, B_6 AND B_{12}. Although I might be tempted, for the sake of efficiency, to just use a B complex or multivitamin, I don't. Vitamins are powerful and potentially intrusive. And don't get me started on B_3. (On a sort of side note, I have found that taking the stomach medication Zantac before my first drink prevents symptoms of alcohol flushing.)

MILK THISTLE. I have come to believe that this is what ultimately saved Tom and me from the full effects of drinking twelve pints in twelve pubs—or whatever the final count may have been. Made into healing elixirs and salves for two thousand years by druids, swamis, Wiccans and naturopaths, it may still be the most powerful hangover remedy available in England, and certainly at Boots.

N-ACETYLCYSTEINE (NAC). I've probably said enough about this, at least for the time being. I think it is definitely the key ingredient. And I've found that I need to take a higher dose than recommended in order to get the desired result, usually at least 1000 milligrams. My bottle says the dosage is 500 milligrams a day. Maybe I'm reckless to take this much NAC. The official health recommendations warn that NAC should be taken with meals, can cause nausea, vomiting or allergic reaction. They also say don't take it if I'm pregnant, breast feeding or have cystinuria. Or if I'm on antibiotics or nitroglycerin. So far, so good!

FRANKINCENSE (*BOSWELLIA*). Although this is the second-most important ingredient, I sometimes trade it out for another anti-inflammatory analgesic, such as CBD oil or perhaps even Chinese snake oil. My preference is for something natural that won't tax the liver. And it needs to be potent.

I've found that I want the maximum recommended dose for all of these things—other than the NAC, which I take as above. And I don't skimp on the analgesic; preventing inflammation is vital. All these ingredients work out to between six and ten pills or capsules, depending on the doses and whether I combine any of the B vitamins. I'm not one of those people who balks at swallowing pills. I prefer it to when my organs go rigid, bowels spasm, brain starts burning, mouth fills with ash and everything I swallowed comes writhing back up.

I also understand that consumers are more attracted to the idea of glugging little bottles than taking big pills; every marketing wizard and their mother has told me that. And I'm fully aware of our deep desire to drink things; it's what started all this in the first place. But I am not in the business of marketing a product. At least, not yet. I'm merely trying to tackle a mystery of the universe.

I find that the timing with which I take the pills matters greatly—but not in a stressful, finicky way. It's simply this: I have to take them all, at about the same time, between drunkenness and sleep. Usually, that is after my last drink and before I pass out. But it is sometimes a couple hours in either direction. An earlier moment, when I'm partway drunk and I'm thinking about the pills, and might not think of them again, I tell myself: take them now! Or if I've passed out for a very short time, then woken up again—take them now! After that, I'm past the point of it working. And into the realm of other things.

Now, if I've taken them correctly, after drinking a helluva lot, I might still wake up dry-mouthed and creaky and worry that it hasn't worked. But I just drink a glass of water, rise up slowly, start to move around, and behold: I may still be tired, even exhausted, but otherwise things tend to be right as rain.

AFTER THE FLOOD (AND INTO THE HEART OF IT)

Dr. Mignonne Mary has been nodding and smiling. And when I finish the lowdown, she pats my arm, just below the catheter. "You're talking about a *lot* of NAC," she says.

"I think it's the key," I tell her. "Any product that's used it—they never give amounts. But I believe it takes this much."

"You can get rid of the milk thistle—so that's one less pill."

"You don't like the milk thistle?"

"I love it," she says. "There's a reason it's been used for these kind of things for so long. But with all that NAC, it should be redundant. The efficacy of milk thistle in these regards is because it makes glutathion."

It kind of blows my mind. After all this time, I still didn't realize this: why, in all my trials and experiments, milk thistle ended up in the column of things that did the job—and why, now, I can put it aside, at least for this concoction.

As for the NAC, she says, "There is no more powerful antioxidant and immune-system rebuilder. And it reactivates the vitamin C."

"Is there anything you think I'm missing?"

"Well, definitely magnesium, probably three different ones, including mag threonate, which will cross the blood-brain barrier. And I wouldn't be surprised if it helps with your outstanding issues—the sleeplessness and exhaustion. But keep in mind, just because you've found something that works, doesn't mean it will

work for everyone. For example, the vitamins you've got, are
they methylated?"

"Um . . . no . . . yes? . . . I don't know." I should almost definitely know what she means. "What do you mean?"

"A lot of people have a genetic variant whereby they can't
break down B vitamins. If it's methylated, it's already broken
down, so they can absorb it."

"This is just what I need to hear," I tell her. My hangover is
feeling a lot better, and I am diligently taking notes. "Is there
anything else?"

"Well, you've got B_1, B_6 and B_{12}, which are good—but what
about B_3?"

"Niacin," I say, or more like spit out. "It's not for me."

Dr. Mignonne Mary gloms right onto that. "There's this
thing," she says, bright eyes flashing. "NAD—it's a niacin derivative. This guy William Hitt came up with it in the '50s—did
some amazing research, down in Tijuana." For all her pristine, soccer-mom prettiness, Dr. Mignonne Mary's appreciation
for underground, renegade brilliance rises quickly to the surface. "He used it to detox people off alcohol, for PTSD, anxiety, depression . . . and now you know what? They did some
tests at MIT, just recently, and found that giving a drunk subject
NAD through IV actually cut his blood alcohol concentration
in half, within minutes. They stopped and started the drip three
times, with recorded results. The stuff was straight up breaking down alcohol and shuttling it out. It actually reduced the
BAC! That is just amazing. And it also improves mitochondrial
function. It actually grows mitochondria! I'm working at getting some for the IVs, but it's still extremely expensive. It is the
future of detox, though. I'm quite sure of that. And it might be
the key to your remaining symptoms—that and the mag. Oh,
and the NAD won't make you red and itchy. What I believe

it would do, though, if you take it enough times, is effectively reroute your brain to not crave alcohol anymore. Wouldn't that be an interesting way to treat hangovers?"

"Um . . . yeah. But why have I never heard of this?"

It is my job to be wary. And I'm also aware that the most effective charlatans these days don't have waxy moustaches, pool-shark nicknames and snake-oil spiels; they have perfect teeth, names like Gwyneth, or possibly Mignonne, and tout the miracles of detox, new neural pathways and magnesium.

"Because," says Dr. Mignonne Mary, in all her loveliness and patience, "I only just told you about it."

And that is more than fair enough.

DURING THIS LONG, tumultuous quest, I have gradually become most wary of blind skepticism. The pervasive habit we seem to have fostered, to disbelieve in something as blatantly possible as a hangover cure—without even realizing we're doing it—is puzzling and bewildering to me, and seems to suggest something fundamental. A subconscious need to suffer for intangible sins? Or maybe an evolutionary mechanism capable of overriding logic and the limits of imagination, so that we don't all get stupid drunk every single day and make a mess of everything until the reins of power get picked up by the most dangerous of lunatics: teetotaling narcissists with unconscionable hair, tiny hands, a love for lies and the will to build walls. As H.L. Mencken once said, "All of the great villainies of history, from the murder of Abel to the Treaty of Versailles, have been perpetuated by sober men, and chiefly by teetotalers."

Wouldn't it make a poetic sort of sense that my missing magic ingredient might be found in some Tijuana über-niacin—and even more so if a side effect of repeated use was a diminishment of the desire for dangerous drinking in otherwise virtuous souls.

That might, in fact, be a way to cure the hangover without also destroying the world.

A MILLION-DOLLAR IDEA

One of the things I'd still hoped to do down here was find Dr. Jeffrey Wiese—the preeminent hangover researcher who, along with *The Hangover* movie, kicked off so much newfound interest in hangovers with his published study on prickly pear, created the best-selling New Orleans–based product No-Ho, then seemingly vanished (along with his elixir) when Hurricane Katrina hit.

But so far, I have only heard back from Dr. Michael Shlipak—Wiese's co-researcher and writer on several hangover studies. He's informed me that they "both stopped doing interviews about hangovers around 2004 as it was overshadowing our career growth."

I suggested we could talk about that instead—the overshadowing aspect, and what comes when you slip out from under it. What I'm most interested in, after all, is the idea—practically a myth, since Wiese won't talk about it—of a doctor who commits so much of his career to the study of hangovers, develops, markets and sells a uniquely successful product based on his findings, becomes frustrated when it's all anyone wants to talk about, in such superficial ways—then forsakes it all to help his drowning city. It can't but sound like the stuff of legends—ones I'd like to hear from him.

For Dr. Shlipak's part, he admits it "would be entertaining to have a three-way call with JW." But still the silence from JW is deafening.

ON THE OTHER HAND, I have finally found Dr. David Nutt—
not in New Orleans, but in the Hammersmith Hospital in London,
where the former British drug czar is apparently hiding out. He's
agreed to have a talk with me, across the ocean, through the
magic of Skype.

"It's good to finally meet you," I say, once the connection is
made—a surprisingly cheery, crimson face filling my laptop screen.

"Well, in hopes of being able to say the same, do you have an
extra million dollars?"

"Sorry. I don't."

"Ah, well. I'm asking everyone. Let's just continue then . . ."

I could say Dr. Nutt lives up to his name, but that is far
too easy. And he is too beguiling to just be nutty, too sharp
to just be whimsical, too blunt to just be enigmatic. No wonder the press has no idea what to do with him. Combine
Nutt's esoteric personality with humanity's persistently inept
approach to hangovers, and the capacity for misinformation
is limitless. For years, newspapers have been reporting Dr.
Nutt's latest claims as such: he's creating a drug that gets you
drunk, another that makes you sober, and that'll be the end of
hangovers.

But according to Dr. Nutt, they've got it wrong. There is, in fact,
no intrinsic connection between his two supposed breakthroughs.
"Except," he says with a twinkling smirk, "they are, both of them,
pure genius." And somehow I'm inclined to believe him.

He's named one Chaperone, the other Alcosynth, and both
are meant to make drunkenness safer, but in very different ways.

"It's important to understand," says Nutt, "that alcohol toxicity isn't linear. If you have three drinks a day, you increase the
risk of dying from alcohol-related disorders fivefold, whereas
with six drinks a day, it is twentyfold. So, really, we want people

to peak at about three drinks." The idea is that taking Chaperone will, in effect, make three drinks feel more like six—thus reducing the amount of alcohol, and toxicity, required for basic inebriation. "The physical effect of alcohol wouldn't be much changed, but you'd get more pleasure from it, more satisfaction. Which would moderate the craving for more."

Whether or not such an approach could work for the orally fixated, fully consumed alcoholic is unclear, but perhaps that's where Alcosynth comes in.

"This is the one that's really mind-blowing. Chaperone's not so radical, but Alcosynth is." As Nutt sees it, the biggest problem with booze, beside its toxicity, is that it doesn't max out: the more you take, the more effect there is, until eventually it will just kill you. People drink themselves to death all the time. The purpose of Alcosynth is to replace alcohol with a drug that has a peak. These are called partial agonists.

"They're a good kind of drug, because they won't kill you. That is how morphine works—as a heroin replacement. And there's one for nicotine too, much less well known. With Alcosynth, the effect builds up through the first few doses, then just levels out at that sweet spot, and it will never get bigger. You could never get so intoxicated that you would fall over or get aggressive or not remember something or vomit, anything like that. And you'd never get a hangover."

Of course, this sounds too good to be true.

"Well, it's not like it was easy," says Nutt. "Alcohol is just about the most complex, most promiscuous drug there is. It accesses different receptor sites in different doses. A low dose effects gamma, maybe dopamine. More of a dose, and you're talking about serotonin and glutamate. So to replicate alcohol, you're looking for a drug that has such a complicated pharmacology."

"And you found one?"

"We found many," says Nutt with a wink.

I have read that Alcosynth is a benzodiazepine derivative. In fact, the three most recent articles all say this. But Nutt is incredulous. "A benzo derivative? I am not going to tell you what it is. But it is certainly *not* a benzo derivative."

I feel bad that his scorn is lost on me. But then, my admission of this seems to be what somehow spills the beans, at least some of them: "Okay," sighs Nutt, "it's a positive allosteric modulator of the GABA-A receptor. We've got some other ideas too. But I'm not going to give it all away. This may not be the final one, but it's the one we're working on."

"And how far away do you think you are?"

"About a million dollars."

"Right. Which I can't really help you with . . ."

But there is something else that Nutt has said that is still sticking out to me. "Glutamate," I say. "Do you know about something called glutamate rebound?"

I came across this term at the very start of my research. And here, at the end, I think it might be the explanation for what I see as the lingering secondary symptoms not yet resolved by my hangover cure: sleep disturbance, anxiety and exhaustion. But now I can't figure out where I originally found it.

"Glutamate rebound is more about long-term, heavy drinking," says Dr. Nutt. "But the process begins right after that three-drink point. That's when alcohol starts to block glutamate receptors. Glutamate is essential in creating energy—and your body never likes being blocked, so after enough repeated drunkenness, day after day, the receptors start to grow in number, to balance things out. But then, if you suddenly stop drinking, there'll be excessive glutamate—which causes all sorts of hyperactivity, restlessness, anxiety . . . so that's part of acute withdrawal. But I'm not sure about your everyday hangover."

So, although glutamate rebound could, by now, be a problem for me personally, it probably isn't for the average drinker—in regards to getting sleep and feeling rested.

"That's more like a . . . let's call it a *GABA* rebound," says Nutt, as if now, together, we're coining new terms. "Since alcohol's a sedative, it can put you to sleep. And after enough of it, you can go into almost *too* deep a sleep. But then the withdrawal starts to fragment that GABA effect, and suddenly you wake up at 6 a.m. and can't go down again, because your brain is hyperactive, in compensation mode."

"Any idea what I can do about that?"

"Maybe not drink so much?" says Nutt, with a cheeky kind of shrug.

"Or if you hurry up, I could drink Alcosynth instead."

"That would work."

"Does it *really* feel the same?" I say, pointing a finger towards the screen. "The same feeling of intoxication?"

Dr. Nutt nods. "Affirmative. I mean, we haven't tried to quantify precisely how similar it feels to alcohol. But yes, it does. There's no question about that. It's similar enough that most people would not be able to tell it apart. Well, maybe a very sophisticated drinker would."

I am unsure of what this means. A connoisseur not of Scotch, champagne or pilsner, but the different ways it feels to get drunk? If so, I might just be sophisticated.

"Maybe," I tell Dr. Nutt, "I could be of some help to you after all."

AFTER THE FLOOD (THE CURSE OF MY CURE)

Getting drunk in New Orleans feels like it did when I was young. Like I want it to. Like it should. It feels cool and adventurous, as if anything can happen—as if the drinks are made of ambiguous molecules, promiscuous compounds, magic potions. It is a bluesy, shimmering neverland with a thousand shakers shaking, brilliantly balanced cocktails, fountains of absinthe, champagne pyramids and hurricanes in glowing beakers, and the night is full of music.

And waking up in New Orleans feels metaphysically right—mornings not so hostile and hollow, but rather big and easy. Moving through the streets is like being inside some intrinsic, ingenious design: all these gothic courtyard fountains and shaded, shimmering pools; all this ivy-laden lattice and weeping willows, filtering and dappling the sun; all the sliding, bending, deeply joyful, intensely sad Dixieland jazz winding through the streets; all this long-perfected local fare, beignets and mimosas, shrimp and grits, oysters and okra and a hundred delicious hairs of the dog—as though it all just is, without even trying, the biggest, easiest, most beautiful place to ever be hungover, so that the hangover itself feels not like a disorder, but instead the natural state of things.

On my final day, I leave the French Quarter and walk to where the streets are straighter, wider, the houses both newer and more rundown. I cross a boulevard and a graveyard and into the Tremé. On a lonesome, nondescript corner, in a lime-green, concrete box, is the voodoo church and shop of Priestess Mambo Marie—one you won't find on the tourist maps.

During my time here, I have been to other voodoo practitioners. Some were unsettling, some derisive, and one truly cruel—I watched as she tormented and ridiculed a girl who had just lost someone close to her. But Mambo Marie is warm and whimsical, her laugh like a tropical storm. And she's the only voodoo priest-ess who's had anything intriguing to say in reference to hang-overs—something about a certain tick. But she was closing up shop and told me to come back today.

Her place has none of the kitschy, claustrophobic menace of so many of the French Quarter emporiums. It is full of weird-ness and mystery, for sure—but with much more space and light. "You again," she says, as I approach the long, glass-case counter. She's seated on a stool, her chin raised, unmoving, as a woman stands behind her, weaving Mambo Marie's hair into a series of braids. I ask if I can record her voice, and also take a photo.

"The voice is fine, but no way about the picture. Can't you see I'm only halfway to looking good?" Her smile is intoxicating.

I turn on my recorder. "So, you said, Ms. Marie, that you know of a cure for hangovers."

"Ah, yes. Okay. Are you ready to hear it?"

"I believe that I am."

"Okay, then. First there is a herb. They call it *bois sur bois*."

"Wood over wood."

"Yes, yes. It is a thick vine that grows over a tree. So you make it into a liquor. And then you take a tick. It is a Haitian tick; I do not know the name. But I know the tick. And it has to be alive—which is why I cannot get it sent here. It's not allowed, you understand."

"Sure. Okay."

"You submerge the tick, alive, in *bois sur bois* and very dark rum. And so you have the drink. So you give it to the drinker—the *drinker*, you understand? He's so much of a drunkard, an alco-holic, that it makes him happy. It is like giving him money, and so he takes it. And then. And then. He starts to vomit. And after that, it gets worse. I have seen it. It gets very much worse. So *sick* by the end of it, he never wants to drink again. And he never does."

"Well, that's a helluva thing," I say, and damned if I don't mean it.

Obviously, voodoo priestess Mambo Marie has no interest in solving hangovers, but rather—like so many oracles, scholars and healers dating back to Pliny—sees the question as an opportunity for teaching, through intense aversion therapy, how painful things can get.

But I've already learned quite a lot about that, and even more so recently. Just last month, having gone so long, drinking and drunk, without physical hangovers but with so many other gathering repercussions, I went to a well-respected shaman to take ayahuasca—an ancient, powerful hallucinogenic. It is said to have complex healing powers that often work—at least at first—by making you feel unimaginably sick. Friends had described it as "the worst hangover ever" and "going through hell, then coming out smarter."

For me, the eight hours between the ayahuasca elixir kicking in and my coming out of it were perfectly terrifying. In a large, dark room, barely lit by a flickering candle, the shaman started singing—some bizarre, wordless, swooping song—strumming and thumping invisible instruments as I began to vomit. But it wasn't reverse peristalsis. Instead, what came up was from way down deep—a thick, black substance, like the liquefaction of long-buried beasts. It just kept coming, and then something worse, more painful, something invisible pulled from deeper still, up through my guts, heart and other organs as I writhed and wretched in the darkness, on and on, until suddenly I could see a clear, visceral vision of three slender, gun-metal-helmeted aliens with glistening scythes. They moved steadily and smoothly, dispassionately, reaping all I'd ever had inside with long, even cuts through my body and brain. I felt every bit of it, yelling and spewing and crying, because of everything I'd ever done and all

I'd ever taken in. They cut through things in my mind I thought were gone, or I'd tried to lose, and peeled them back so that other screaming parts of my brain could see, then flicked them with their blades, spinning into the darkness . . . until, finally, it was over. The long night was done, and I walked into the day.

Before I leave Mambo Marie, I ask her one last, immensely stupid question: "I've heard of voodoo practitioners trying to take someone's hangover and put it on someone else. Have you ever heard of that?"

"What rubbish!" she says, waving her hands, disrupting her braids. "But that is what people do, is it not? They try all kinds of rubbish, until the day they stop. And of course, one way or another, one day or another, everyone finally stops."

A KIND OF CONCLUSION

There are many good things about hangovers. Despite what some researchers suggest, they often, and undeniably, act as a formidable disincentive to getting stupid drunk. True, for some, there can be a reverse effect, and the hairs of the dog become a monkey on your back. But even that can sometimes save you by bringing you to your knees.

In certain ways, the hangover is akin to our ability to feel pain. We know, at least theoretically, it is a warning system, so if we fall asleep with our foot in the fire, we'll pull it out before we go up in flames. But when you're deep inside an awful hangover, it feels like so much overkill, torture, so hard to reconcile—does it *really* have to be so extreme? And why must it go on for so goddamn long, well after the warning's been heeded?

But really, in evolutionary terms, the persistence—possibly even the intensification—of our susceptibility to hangovers makes sense. Sure, moderate drinking can bring disparate groups together, increase the birth rate, fight off certain diseases, spark new connections, new ideas and new art, and increase the enjoyment of life. But too much can do just the opposite—and then people are dying in the streets, babies aren't being born, and the basic structure of society is falling into ruins.

The near-universal awareness of hangovers does appear, at least in the long run, to regulate the balance of human and booze. And what's a bit of personal pain compared to the survival of the species? It could also explain why we're so resistant—on a possibly innate, seemingly illogical level—to the idea of a simple hangover cure. I've come to believe this is true—not only from looking at histories, reading a bizarre array of books and talking to doctors, philosophers, psychologists and various other people, but more so from my own experience.

Having discovered the antidote, I took it and took it and took it and, in so doing, caught a long, unsteady glimpse of what consuming immense amounts of alcohol with no obvious physical repercussions can quickly become. I went through an intense and dramatic descent—from sunward arc to rocky crash, unbound potential to slobbering beast. And I learned that if you remove the most aggressive and acutely physical symptoms of hangover but leave the more insidious ones—exhaustion, lethargy, anxiety, hollowness, depression—it's like slipping into an alternate universe, one full of fears and problems you never knew existed, your life and liver constrained by scars and bars.

So now, even as I work to improve the recipe—adding different kinds of magnesium and methylated forms of B vitamins and learning about NAD—I'm not sure I'm working for the greater good. And there also may be simpler reasons to not solve the problem at all.

I recently met a fellow freelance writer for dinner. And pretty soon we were talking about hangovers. Instead of the standard complaints, she dialed in on the surprising benefits—in almost precisely the same terms as Dr. Richard Stephens had, about those Saturday morning jam sessions in Liverpool. But the example my very professional and talented colleague gave involved the more nuanced art of recording outgoing messages: "There's just something about it. My voice is more chill and scratchy, and the message comes out more groovy, like *'Heeeyyyy, you've reached Sarah's number, you know what to do and when to do it . . .'* I like the *sound* of hungover me. I mean, obviously being hungover is the worst, but it has a useful side. It's like an altered dimension where I'm a mellower, more relaxed version of myself. Like there's only so much I can worry about." To me this sounds like the Shining mixed with the Paradox of Choice—and it could be a key to all sorts of other issues.

According to *Business Insider UK*, the London-based music-ticketing app Dice has recently introduced "hangover days" so employees can skip work without repercussion if they've been out too late at a gig. Says Dice CEO Phil Hutcheon, "We trust each other and want people to be open . . . There is no need for a fake sick bug." And they might be on to something: not just avoiding morning-after screwups at work, but coaxing the hangover out of the shadows—embracing it, even. It seems like a good idea, at least for certain jobs. But then again, think of those Secret Service agents protecting JFK. What, in the end, might have been worse: to be or not be there—fighting the good fight badly, or seeing it all happen on a hotel TV, inept and screaming for taking a "hangover day"?

As with everything else, over all these years, I just don't know. Are hangovers good or bad? Are they affecting us worse than ever, but for vital reasons—an evolutionary necessity, or just a leftover, meaningless scourge? Is there a legitimate reason we fake all this know-how even while throwing up our hands, throwing back so many drinks, throwing up beside our beds? Or do we simply, deep down, not give a damn about illness until all the way inside it?

Maybe an all-consuming crapulence, taking hold for just one day, allays our most buried fears—a sort of psychological inoculation against the inevitable sickness, breakdown and darkness from which we can't return: a preparation for death more potent even than sex or sleep. Who the hell knows? But I do believe hangovers can lead to bigger things, in small but sometimes helpful ways.

Just recently, in my homeland of Canada, our national news gave us this perfect human-interest headline: "Hungover Customer Brings Heaps of Business to Struggling Alberta Fish and Chip Shop." The story went something like this:

A man named Colin Ross was drifting on foot through the streets of Lethbridge, Alberta, in that way you sometimes do when you drank a lot the night before—both aimless and searching for the right place to be. He came upon a diner at the far end of a parking lot, next to Tim Hortons, where the Black Tomato used to be, but now the sign read WHITBIE'S FISH AND CHIPS, and Ross trundled in.

He ordered the three-piece halibut and fries, which came out golden and steaming. And it worked like magic. As Danielle Nerman, writing for the CBC, put it, "Ross devoured his meal and his hangover began to dull. As clarity set in, he realized the shop was empty."

So, as sometimes happens when you're becoming human again, our man Colin started talking to the only other person around. That was John McMillan—a classy, old-school guy who, a few months shy of seventy, had opened this clean, cool place with fish and chips so good they'd curbed Ross's hangover. Yet it seemed nobody even knew this place was here, tucked in the corner of a parking lot. McMillan confessed that it was always empty, and he couldn't even pay himself.

So, as sometimes happens when you're coming off a hangover and want to do something good—even if it's just with your thumbs and your phone—Ross took a picture of the place, wrote a quick little paragraph about this stand-up gentleman who had helped his heavy hangover, then suggested that everyone he knew should try this good guy's fish and chips. And Ross knew a lot of people, who knew a lot of others.

So now, if you're hungover in Alberta and in need of golden, flaky healing, you might have to stand in a line that stretches through the parking lot. But apparently the wait is worth it. Also, in the photo accompanying the story, I couldn't help but notice, standing on that Canadian counter, next to the shining plate of Whitbie's fish and chips, a genuine frosted bottle of Scottish Irn-Bru. And so I might also report: Our man Dart, it appears, has completed another mission.

IN OTHER RELEVANT news are three strangely similar criminal cases: all recent, all occurring within blocks of where I live and all pertaining to the most profound levels of hangover. In fact, each involves an act of sudden deviance, a state of intense reckoning and, finally, contrition.

The first began on October 4, 2016, with what CBC writer Andre Mayer calls "The Lob Heard Around the World"—a beer can tossed from the stands at Baltimore Orioles outfielder Hyun-soo Kim during the seventh inning of an extremely tense one-game playoff against the Toronto Blue Jays. Somehow Kim retained his focus enough to make the catch as the can fell like a bullet casing on the turf behind him.

But even with so many cameras covering the action, it was unclear who had thrown the can. The search for a culprit, dubbed "The Beer Tosser," began immediately, as did an outpouring of vitriol, astonishment and derision.

By ten o'clock the next morning—when Ken Pagan came to consciousness on his friend's couch, covered in hazy dreams and dread—the story had made every major news outlet, rewards had

been offered for information, and even Stephen King was in dis-
belief, twittering, "Hey, whatever happened to polite Canadians?"

Ironically, Pagan seemed to be exactly that: and not just a
soft-spoken, thoughtful, generous, quintessentially polite Canadian, but one with such reverence, love, aptitude for and understanding of sport, particularly baseball and hockey, that he might
have been a pro athlete if it weren't for his other abilities and
interests vis-à-vis the human condition and how we learn about
it. Employed as a journalist by the very media company now
offering a $1,000 reward for information about the despicable
Beer Tosser, one could say that Pagan—waking on the couch that
morning—had been sleeping with the monkeys, and was now in
their clutches.

While acknowledging that he was drunk, Pagan can still only
explain his moment of immense idiocy in bewildered Canadian
terms. "Honestly, if I was to break it down blow by blow, I'd be
speculating," he recently told journalist Andre Mayer. "It was an
impulse. . . . I equate it to if you've ever taken a bad penalty in
hockey and realized, 'What did I just do?'"

What he'd done was affect the game of baseball, the way beer
is sold in stadiums and almost every facet of his future.

In the end, it was the disproportion between the infraction
and the consequence—the moment of stupid drunkenness and
still-ongoing metaphysical hangover—that saved Pagan, at least
legally. While recognizing the severity of his action, the judge also
acknowledged the degree to which the Beer Tosser had already
suffered. He'd lost the job he loved, his sense of self, his career
and his pride, and he had been publicly shamed on social media—
the town stocks of Facebook, the drunkard's cloak of Twitter.
So the judge let Pagan go, but with conditions, probation and a
world of monkeys on his back.

An even more public Toronto hangover began shortly after

the bars closed on April 26, 2017. Marisa Lazo was drunk and alone and feeling reckless . . . and that's when she saw the construction crane, looming thirty stories above a half-built condominium tower.

Though requiring considerable more effort than the stupid toss of a beer can, it was still just a matter of intoxicated impulse—the sudden drive to feel more alive—that sent the twenty-three-year-old Lazo over the chain-link fence and up that giant crane.

It wasn't easy, but she kept on climbing until she was all the way up—at the very top of this iron colossus. And, man, it was even better than she'd thought. Standing in the dark, swirling air, above the bright, brilliant, stretching world, she might have sounded her barbaric yawp. At the very least, she took some selfies, then eventually turned to climb back down. And that's when Marisa Lazo—four hundred feet above the street—slipped. And fell.

That day, the sun rose on downtown Toronto at 6:17 a.m., and it revealed the tiny shape of a woman on a small, dangling perch in the sky—so high up it was like picking out a swallow from the clouds.

By early rush hour, the whole city was gazing upward—necks craning from cordoned-off sidewalks; commuters pulling over, their way blocked; radio announcers baffled; emergency workers scrambling.

But despite the inner-city hassle and the piling-on of trolls, the story of Crane Girl's very public hangover somehow unfolded inversely to that of Pagan the Beer Tosser. A long-scope photo of Lazo, mysteriously suspended, somehow sitting on a tiny bench in the air, appeared on social media, in the palms of people's hands. And it was undeniably romantic—an image of urban ennui and dangerous solitude, her long hair twisting in a different air current.

The photo that had surfaced of Pagan in the stands—taken right after the infamous beer toss and released to the public late the next day—had shown an expression familiar to anyone: that of a guilty kid trying to become invisible. And the general response to these two pictures couldn't have been more different. While the face of the suspected Beer Tosser had ignited fires of unmitigated scorn—as if not one of us could even imagine such a thing—the visage of the far-off Crane Girl became an instant, empathetic meme—a self-deprecating shorthand for how we all feel sometimes.

And in stark contrast to the bounty-fueled manhunt for Pagan, Lazo's predicament became a real-time rescue mission: with all the pathos of a child trapped in a well, but even more suspenseful, perilous and visible. The two-hour rescue was breathtaking. And then the fireman who'd pulled it off, holding her in his arms as they descended to the earth, was also great at sound bites, quipping to reporters that, since she was now safe, his main concern was getting to the hockey rink in time—his beer league team was in the playoffs and he was the starting goalie. The nation rejoiced—though somewhat chagrined to see Crane Girl in cuffs, facing several charges of criminal mischief.

About a year later, the case went to trial and Lazo pled guilty. She offered a heartfelt apology and did her best to explain what had happened. Her description made me think of the Stratosphere, and that all-of-a-sudden reboot. But of course, when Lazo fell, she wasn't attached to anything. Still, she managed, in midair, to grab hold of a cable—which she then slid down at breakneck speed, a hundred feet or more, somehow landing on a small, swinging platform.

The rush of adrenaline through her body would have been as much as it could ever be, almost certainly triggering the phenomenon of quick-sobering. Then, as she dangled there, in the middle of the air, the adrenaline would have kept on pumping, until there was none of it left and nothing she could do, the world below turning from night to day. It was as if, hanging there over the city, she had found a heretofore-undiscovered root, or perfect brand-new definition, of being . . . *hungover.*

There were moments when Lazo—increasingly despondent and desperate—thought about jumping but decided, "I can't do that to myself or my family." In court, it came out that she had been abused as a child and had two severely disabled younger siblings, for whose care she would eventually become responsible.

In pronouncing his verdict, Justice Richard Blouin acknowledged Lazo's dangerous, drunken disregard for public safety, but also the surprising twists of fate that sometimes allow for grace: "Ms. Lazo was obviously in a very dark place when this happened," he said. "She's done an amazing job of climbing out of that dark place. In a very unusual turn of events, this event has maybe opened up some things in her life that were kept back and not dealt with."

So while Lazo pled guilty, saying that "climbing that crane was a terrible idea" and something that would always haunt her, the official terrestrial judgment was one of redemption— and also absolute discharge, on each and every count. It was shouted from the rooftops: *"The Crane Girl is free! We are the Crane Girl!"*

Finally, there is Homeless Jesus, and his mysterious disappearance. This one contains echoes of both the Beer Tosser and Crane Girl, at least in spirit. Though the events transpired a while ago now, they are still ongoing in certain ways. And it is also the case closest to me, both metaphysically and geographically.

Two blocks from where I live, in downtown Toronto, is a beautiful old church, surrounded by thick, gnarled oak trees. St. Stephen-in-the-Fields is exactly equidistant between the nightclub I used to manage and the bar I used to own. Across the street is the fire hall, and halfway down the block is the daycare where my son, Zev, used to spend part of his early days.

During that time, a heavy dark-bronze statue was set down in the earth between a corner of the church and the accompanying sidewalk. It was the figure of a cloaked man, sitting on the ground, his hand held out as if begging for change. And if you looked at it more closely, his palm was cut wide open. Though originally entitled *Whatsoever You Do* by its sculptor, Timothy Schmalz, the bronze beggar became known as Homeless Jesus.

Whenever we'd pass by—almost every day, Zev on my shoulders or holding my hand—we stopped so he could leave some coins, fitting them into the hard, yet flesh-like, stigmata. Every time we came back, the coins had disappeared, and Zev liked to guess where they'd gone—what cool little things someone had used them to buy.

Then one day, along with the coins, Homeless Jesus disappeared too. Gone without a clue. And it was hard to know which was more strange: why someone would do such a thing, or *how*— steal a statue so heavy, in so many ways.

People shook their heads, and wondered. Zev had a hundred guesses. And knowing most of the baffling rogues on these streets, I had a few of my own.

Then, four days later, Homeless Jesus miraculously reappeared, along with a handwritten note. Though tricky to read from being smudged in the rain, the words appeared those of a deeply hungover soul:

I'm sorry. It seemed like a good idea at the time.

AS FOR ME, it's been a while since New Orleans. And I still haven't finished this book. To quote a jackass friend of mine, from some time ago now, "Of course you're still writing that thing. As soon as you stop, you'll be out of excuses. Or at least they'll be harder to sell." And he didn't just mean about drinking.

I recently got the results of the test kit I sent to the British/American Gut Study. They arrived as a series of comparison graphs and pie charts without much guidance or context. I didn't really know what I was looking at—but even so, it looked a bit off.

From what little I could comprehend, my collection of microbes was very different, in a few specific ways, from that of the average gut. I appeared to have none at all of two common types of bacterium, and then twenty-five times the norm of another. But I had no clue what this might mean. So I sent the data to Dr. Timothy Spector, of the mice and cheese, who had first put me on to the study, and asked if he'd take a look.

His assessment was brief, vague and somewhat daunting. He referred to my microbiome as "low diverse, unhealthy looking," said that it meant I am "more susceptible to disease but not necessarily ill," and then signed off with what you never want to hear from a medical doctor: "Sorry. Best, Tim."

I don't know what to think. "Not necessarily ill" is a tricky prognosis—especially in light of how I feel these days. I don't feel well. I am filled right up and weighed right down—so that sometimes it's hard to sit, or stand, or walk across the street. And I hate this feeling I can't seem to shake—that now every day is just another morning after. Never a new beginning.

In hopes of better health, I have retreated to Vancouver to spend some time with my folks and visit various doctors. After yet another appointment, I cross the street to catch a bus. But next to the stop is a small antiquarian bookshop. So I go in there instead.

It is how you might imagine: quiet, with books in rows on shelves and in stacks from floor to ceiling. At the front of the shop, behind the large wooden desk that serves as a counter, stands a woman with long silver hair, magnificent lines in her face and perfect posture. She is at least two decades older than me. There are three books open on the desk in front of her, and we exchange nods as I move into the stacks.

I spend a while looking at titles, running my hands over the shelves. I have no place to go and might just keep on going, touching every book. The woman and I glance at each other until finally she asks, "What are you looking for?"

"Books about booze," I say. "And drinking. And getting drunk. And hangovers."

And then she says something that stops my hand.

"I love hangovers."

"What's that?" I turn and walk toward her.

"I love them."

In a photo, she might look elderly. She is calmly guarded, but also electric. It is like seeing an old soul that has somehow grown young as the young body has somehow grown old. She is very beautiful.

"Can you tell me why?" I ask her.

She nods. "I can drink almost anyone under the table. And then I get extremely hungover, sometimes for days. But the thing is, I love it. I love it so much more than getting drunk."

I suggest to her that this might have to do with choices, and not being required to make any. She asks me to elaborate.

"It's like you've put your body in a state of crisis. But you know it's one with a time limit. So your mind gets a brief vacation—because this is what you have to deal with, and for now there's nothing else.

"That's exactly it." Her eyes flash. And we start to talk.

Part of me is aware of the strangeness of all this—or, rather, the extreme narrative convention. It's a dust mote-filled antiquarian bookshop, after all, where the silver-haired bookseller speaks like a prophet about the one thing to which I'm still connected, cannot lose.

But there's another part of me that remembers how it used to be—when I was young, lean and thirsty, and every day rolled out like scenes from a film: interconnected, intoxicating and stretching into the next.

She likes what I said about time limits. It is, she explains, how she's been living her life—in three-year reinventions—based on reading Kierkegaard and his ideas about meaningful change. So, every three years, she embarks on a new journey. The last few cycles were Buddhism, then bars, and now she is partway through Quakerism: "It is my way of traveling to all sorts of places, for a good length of time, but without having to go very far."

I ask if she minds if I take some notes. She wants to know why, and I tell her I'm writing a book about hangovers. She looks at me steadily, and we smile. I ask her name, and she says that she'll tell me, but I can't write it in my book or tell anyone else.

"What about my publisher, just for fact-checking?" I ask. "In case you say something truly unbelievable?"

She gives me a placating nod. I write down her name, then ask about the years of bars.

"I'd go to a different one every day, starting at happy hour. I'd drink and talk to the people there, and listen. A lot of the time, I went to the Legions. I've always done that, actually. My parents were in the military. I was a child of the war, and I felt connected to those old men. A lot of them gave their lives, even though they survived, and alcohol was their only retirement plan. The bar stools were stained with urine. That's where I fell in love with hangovers."

She watches my hand as I write.

"Of course, there are still a lot of people and places like that. And I still go see them and have a drink. That's one of the things about these three-year journeys: I can take a bit of each into the next. I still practice Buddhism, I still go to a bar once a week or so, but now I'm doing this: I'm a Quaker who works in a bookstore—what you need to read is *John Barleycorn,* by Jack London."

This advice is offered without a pause, as if she knows I should hear it instantly. It is not only urgent and perfect, but also presumptuous. After all, *John Barleycorn* is a classic. And if I'm really doing what I say I am, I'd be a fool to not have read it. But of course, I am a fool. And she just knows.

"It is everything that a story about drinking should be: mistaken, flawed, true and revelatory—full of beautiful words and utterly terrifying." We are moving through the stacks together now, looking for the book. I would probably follow her anywhere.

"I have always loved alcohol," she says. "But, oh, you have to be so careful with it. What Jack London learned is that when you mess with something so strong—get right in and roll around—you are inviting death into your bed. And then you can't just kick it out."

I get a picture in my head of how I sleep now—so often alone, fully clothed, waking in starts and drenched in sweat, as if there are specters in the bed: on one side a cure, on the other a curse, with just a single serpentine *S* between.

She finds the book—a slim, weathered paperback—and I tell her I'd like to buy it.

"When you're done," she says, as I complete the payment, "come back and we'll have a drink in the shop. Sometimes, at the end of the day, I open a bottle of wine."

I say that I will, and she can tell me about her hangovers.

She shakes her head. "My hangover stories are boring. Like Buddhism." Then she smiles as she hands me the book. "I like it that way."

So now I know what I'll do today. I'll start reading this book. But as I walk out of the shop, I'm still not sure where—on a city bus, in that sunlit park around the corner or at the bar just down the street.

It is choices like these that decide the morning after. ◄█►

Acknowledgments

If not for Jacqui Bishop, Bob Stall, Jennifer Lambert, Samantha Haywood, Rob Firing, Angela McDonald and Yannick Portebois this book might not exist, or I might not exist, or at least neither one of us would be very good. I am indebted to you, and love you all, in great, essential ways.

The same goes for Mike Wasko, Russell Smith, Jonathan Dart, Brendan Inglis, Anne Perdue, Tabatha Southey, Derek Finkle, Mike Ross, Max Lenderman, Greg Hudson, David Lightfoot, Jeff Lillico, Mike McRobb, Bill Rogers, Lisa Norton, Lee Gowan, Bret Lefler, Craig Applegath, Fraser Beale, Keith Darvell, Marci Denesiuk, Janine Kobylka, Saskia Wolsak, Cassidy and Reilley Bishop-Stall, Josh Stall, Olive Bishop-Gladue, Rosie Trudel and Kyara Innocent. You keep me keeping on.

This book took a helluva long time to write, which is why there are so many people to thank—including those who helped with chapters that didn't make the final cut. But for a detailed breakdown of official acknowledgments and how they relate to specific parts of the published book, please read on into the elaborate "Notes on Sources." If you can't find yourself here, you're probably over there. Or I forgot to put you in and owe you a number of beers.

I need to thank my wonderful family: the Bishops, Stalls and Bishop-Stalls; the McDonalds and all the Greenaways, no matter how extended; as well as the Stall-Paquets, Tessier-Stalls, Rosses, Trudels, Tremblings and Hauers. In the time it took to write this book, we lost several of our beloveds, including Deb, Benoit, Johnny, Josie, Rose and Maria. I wish we could have just one more drink together.

On that note, I really should acknowledge a slew of heroic barkeeps, including Christine Thibodeau, Andrew Di Battista, Sally Gillespie, Karla Cruz, Danny Boy Mojarro, Morley Wilson, Marko Yovanovich, Chris Stevens, Laura Beattie, Phuong Nguyen, Jeff Kennes, Marcus Bankuti, Mairi MacEachern, Lauren Mote, Clinton Pattemore, Hugo Dallaire, Nishan Chandra, Darren Jones, Zoe Will, Zach Wallace, Brian Grant, Josh Drebit, Nils Boese, Paddy Gallagher, Queen T, Princess, Anami Vice and no doubt many others whose names are lost to myth and mist.

Among dozens of experts in various fields, many were incalculably generous, including Brian Kinsman, Adam Rogers, Dr. David Nutt, Dr. Vivian Nutton, Dr. Mignonne Mary, Dr. Jason Burke, Dr. Timothy Spector, Dr. Richard Stephens, Dr. Joris Verster, Dr. Richard Olsen, Dr. James Maskalyk, Dr. Michael Shlipak, Michel Kleiss, Cason Thorsby, Jason Walter, Richard Olson, Zachary Leader, Mike Danielson, Marcel Roeper, Elard Tissot, Bronwen Erickson, Todd Caldecott, Giuliano Giacovazzi, Toby Paramour, Bruce Ewart, Andrew Parr, Richard Wood, Charles Chalcraft, Paul De Campo, Liz Williams, Marita Woywood Crandle and Mambo Marie.

So many friends, loved ones, colleagues and others gave me help and expertise in the form of research, information, invitations, assignments, patience, kindness, gigs and anecdotes. I particularly want to thank Rita Gilly, Jeff Warren, Avril Benoit, Ibi Kaslik, Chris Rodell, David Yarrow, Jill Tomac, Grace O'Connell,

Andrew Elkin, Trevor Cole, Margaret Peterson, Geoff Carter, Tom Dietz, John Lekich, Asia Marrion, Charlie Locke, Joanne Schwartz, Jane Thompson, Maria Trestrail, Claudia Dey, Dave Bidini, Melanie Morassutti, Thom Vernon, Ken Murray, Bert Archer, Lisa Neidrauer, Anne-Marie Metten, Mark Medley, Joy Kogawa, Stephen Andrews, Brodie Brigold, John McLachlin, Nashville Lotus, Sarah Musgrave, Ryan Knighton, Lee Gowan, Jane Thompson, Adrian Trembling, Dre Dee, Robert Hough, Jenny Patterson, Duncan Shields, Evgenia Roza, Lindsey Reddin, Pat Fairbairn, Andy Lyberopolous, Michael Murray, Melissa VanCuren, Bruce LeFevre, Penny Mason, Catherine Jackson, Anthony Abrahams, Phyllis Simon, Mark Sumner, Jennifer CK, Sarah Pomfrey, Jenna Bartlett, Philipp Geissler, Flavia Zaka, Ken Craig, Nick Wasko, Yacine Dottridge, Julien Boumard Coallier, Blair Williams, Charles Francis, Charlie Bossy, Paige Fletcher, Catherine Veteri-Kolybaba, Tina Siegel, Nadia Shahbaz, Anna Van Straubenzee, Cathy Martin, Alyson Soko, Navan Suchak, Matt Edison, James Moyer, Wendy Wilson, Pamela Couture, Flora King, Jena Bechamp, Mark Sanders, Thijs Brundeel, Jason Walter, Brenna Haysom, Benjamin Emile Le Hey, Brett Leppard, Sharlene Fernandes, Kathryn Borel, Justis Haucap, Patricia Westerhoff, Daryl Hogtown, Lucy Seteran, Melanie Buddle, Ariel Ng, Lenni Jabour, John McAleer, Bronwyn Singleton, Gillian Grant, Marina Hasson, Michael McDougal, Aefa Mulholland, Sandra Banman, Sarah Sturgis, Bill Zaget, Emily Sanford, Phillip Preville, Doug Bell, Gary Ross, Mike Danielson, Stephane Beauroy, Alex Snider, Marni Jackson, Chantal Darvell, J.P. D'auteuil, Tonisha Bat, Beth Havers, Liam Wilkinson, Nicola Blazier, Ryan McCann, Rebecca Cohen, Kim Banjac, Trevor Walsh, Bradley Friesen, Susan Greer, Richard Poplak, Ana Buncic, Michael Stein, Robert Perisic, Miriam Toews, Timothy Taylor, John Fraser, Anna Luengo and Laura Catalano.

A particular acknowledgment is due to Irene Spadafora, who never stopped sending me random things about hangovers, even long after the book should have been done.

And then there's Yannick Portbois, whose name has already been mentioned—and who no doubt appears more than any other in the extensive Notes on Sources. Without her overwhelming generosity, perseverance, aptitude and skill, so much of this book would not exist.

I can't overstate my appreciation for David Yarrow and Dennis Schuster, two formidable artists who have donated some of their work to this book. And I am so grateful to the talented Michelle Kath, daughter of the musician and songwriter Terry Kath, who generously gave permission for her father's lyrics to be included here.

Several people went beyond the scope of their jobs to help me with some crucial licensing, including Robi Lubliner at NBCUniversal, Laura Lacey of BMJ Case Reports, Joanne Smith at Figment Films, Larry McCallister at Paramount Pictures, Jochen Schwarz and Kimberly Bianchi of Havas Worldwide, Timothy Lloyd of the *Journal of American Folklore*, Andy Bandit at 20th Century Fox, Shannon Fifer of WB Motion Picture Rights and Claire Weatherhead of Bloomsbury Publishing.

I'd also like to acknowledge and thank the staff and management of the Hospital Club, Weeke Barton Guesthouse, the Glazebrook House Hotel, Rogner Bad Blumau, Hogtown Pub and Oysters, Trinity Common, Supermarket, the Fringe, the Red Room, Hotel Schloss Obermayerhofen, Seehotel Grundlsee, Almdorf Seinerzeit and Sveti Martin, as well as the good people at Glenfiddich, Jägermeister, L'acadie Vineyards, Southbrook Vineyards, Kalala Organic Estate Winery, Resurrection Spirits, SummerGate Winery, Beaumont Family Estate Winery, Jesson + Company, William Grant & Sons, Rooftop Agency, Drinkwel,

Hangover Gone, Sobur, PreToxx, Thomapyrin, Reset, Party Armor, Gaia Garden Herbal Dispensary and Zombie Survival Camp.

I will always be thankful to Tracy Fladl, Nanja Scheurer, Alexsander "Olek" Justynowicz and a few others who can't be named, for helping me out in particularly tricky situations.

Then there are those brave souls (many of whom I dearly love) who let me do experiments on their precious earthly bodies. They include several people already mentioned above, as well as David Stall, Rosie Trudel, Caitlin, Jeremy and Gabriel Stall-Paquette, Youri and Sasha Tessier-Stall, Aaron Krajeski, Brenna Baggs, Annie Gregoire, Cameron Murray, Erik Gaudette, Julien Bournard Coallier, Charles Francis, Tom Avis, Jay Gladue, Toby Berner, Jeff Meadows, Jeff Topham, Masa Takei, Nicole Oguchi, Nick Rockel, Lauren Stope, Migs de Castro, Robin Esrock, Adrienne Matei, Mark Stope, Bill Upward and Randy Baker, with apologies.

I want to give a singular moment of gratitude—the kind I hope can last forever—to Anne Collins, Ernest Hillen, Paul Wilson, Scott Sellers, Gary Shikitani, Amy Hughes, Cathy McRae and my man Paul Quarrington, for helping me learn what it is to be a writer. And my love and devotion go to Dorothy Benny and the wondrous Quarrington gals.

This book and the writing of it owe a great debt of thanks to the Joy Kogawa House, the Toronto Arts Council, the Ontario Arts Council and the Woodcock Foundation, as well as the staff at Lillian H. Smith Library and the Osborne Collection of Early Children's Books.

Everyone at HarperCollins Canada, Penguin Books USA and DuMont has been amazing, with special thanks to Noelle Zitzer, Lloyd Davis, Patrick Nolan and, of course, the brilliant, kind and superhuman Jennifer Lambert. Thanks also to Iris Tupholme, Alan Jones, Lisa Rundle, Michael Guy-Haddock, Cory Beatty,

Melissa Nowakowski, Natalie Meditsky and Lola Landekic at HarperCollins and Matthew Klise, Christopher Smith, Yaima Villarreal and Mary Stone at Penguin Books.

I am always indebted to the whole team at Transatlantic Literary Agency, especially Barbara Miller, Lynn and David Bennett, Shaun Bradley, Stephanie Sinclair and the ever patient and miraculous Samantha Haywood.

As I've now begun to acknowledge some folks twice, I'd like to do so very specifically, and finish by thanking a few incredible people who, at some point in the past few years, swooped in and literally saved me: Yannick Portebois, Mike Ross, Bill Rogers, Jonathan Dart, Tabatha Southey, Craig Applegath, Grace O'Connell, Greg Hudson, Jeff Lillico, Anne Purdue, Brendan Inglis, Angela McDonald and, of course, my mum and dad—who are not only the best parents I can imagine, but also two of the world's best editors. I owe all of you more than I could ever repay.

PERMISSIONS

Quotations from *Everyday Drinking* on pages 15, 39–40, 85, 103, 162 and 317 © Kingsley Amis, 1983, *Everyday Drinking*, Bloomsbury Publishing Inc.

Quotations from the film *The World's End* throughout Part Five (pages 145–175) courtesy of Universal Studios Licensing LLC.

Graph on page 112 courtesy of Google Books Ngram Viewer, http://books.google.com/ngrams.

Illustration on page 129 is Plate 8 from *A Warning-piece to All Drunkards and Health-drinkers Faithfully Collected from the Works of English and Foreign Learned Authors of Good Esteem*, published in 1682 (engraving), English School (17th century)/British Library, London, UK/© British Library Board. All rights reserved/Bridgeman Images.

Dialogue from the film *Die Hard: With a Vengeance* on page 178 © 1995, written by Jonathan Hensleigh. Twentieth Century Fox. All rights reserved.

Illustration on page 200, *The Head Ache*, is a satirical cartoon by George Cruikshank (1792–1878)/Private Collection/Bridgeman Images.

Painting on page 253 by Henri de Toulouse-Lautrec, *The Hangover (Suzanne Valadon)*, 1887–1889, oil on canvas. Harvard Art Museums/Fogg Museum, bequest from the Collection of Maurice Wertheim, Class of 1906, 1951.63. Photo: Imaging Department © President and Fellows of Harvard College.

Photograph on page 255 © David Yarrow. Used with permission.

Painting on page 280, *The Drunken Silenus*, c. 1617–18 (oil on panel), Rubens, Peter Paul (1577–1640)/Alte Pinakothek, Munich, Germany/Bridgeman Images.

Poetry excerpt on page 293 from "Everything," from *The Roominghouse Madrigals: Early Selected Poems 1946–1966* by Charles Bukowski. © 1960, 1962, 1963, 1965, 1968, 1988 by Charles Bukowski. Reprinted by permission of HarperCollins Publishers.

The image on page 311, *CT Scan Skull Knife Blade*, from the article "Retained Knife Blade: An Unusual Cause for Headache Following Massive Alcohol Intake" by O. Lesieur, V. Verrier, B. Lequeux, M. Lempereur and E. Picquenot, in *Emergency Medicine Journal*, provided courtesy of BMJ Case Reports © 2018 BMJ Publishing Group Ltd. All rights reserved.

Notes on Sources

The following notes pertain specifically to passages in the book where a source or quote has not otherwise been referenced, or where some aspect of credit should still be given. They are intended to function cooperatively with the bibliography, to provide as much information as possible about where I got what.

Preface: A Few Words About a Few Words

Dialogue and permissions from *School of Rock* were generously provided by Paramount Pictures, with my sincere appreciation for the efforts of Larry McCallister. Thanks are also due to Detour Films and the great Richard Linklater.

Clement Freud's *Hangovers* is a small, funny book that also proved to be a bountiful, helpful source. Every time I've quoted Freud directly, it is from this book.

Barbara Holland's *Joy of Drinking* is a joy to read, and you'll notice how often I turned to it.

As mentioned in the text, I have not been able to find an original source for the final quote of the preface, so often attributed to Kingsley Amis. Both Zachary Leader, Kingsley Amis's official biographer, and Martin Amis, his official son, seemed to agree that it sounds like him—but if anyone knows better, or for sure, please do let me know.

Welcome to Your Hangover

This introductory chapter is a sort of compendium of much of my research referenced throughout the book, including several conversations with physicians and various health practitioners who were kind and patient enough to answer my many questions. It also owes something to several medical papers referenced in the bibliography, as well as the article "What Is a Hangover, Really?" on StuffYouShouldKnow.com.

The reference to an ancient fish whose jawbone became our inner ear is from Neil Shubin's book *Your Inner Fish: A Journey into the 3.5-Billion-Year History of the*

Human Body, suggested to me for this very purpose by my astute cousin, Adrian Trembling.

PART ONE: WHAT HAPPENS IN VEGAS

It is worth repeating some of what is in the Acknowledgments to credit *Sharp* magazine, and editor Greg Hudson specifically, for everything I was able to do in Vegas and for providing so much support. Greg and the rest of that team were invaluable in the on-the-ground research of this and other parts of the book.

I found out about the *British Medical Journal*'s James Bond Martini Study through Barbara Holland's *Joy of Drinking*.

"The Morning After the Dawn of Time" was created using pieces of information and bits of scholarship from many sources, including Ernest L. Abel's book *Intoxication in Mythology*, Kenneth C. Davis's *Don't Know Much About Mythology* and John Varriano's *Wine: A Cultural History*. My friend Saskia Wolsak put me onto the myth about Enki. This section also contains the first quote from Kingsley Amis's brilliant seminal essay "On Hangovers," most recently published as part of the Amis collection *Everyday Drinking*. It is quoted many times throughout this book, with the permission of Bloomsbury UK.

"Dionysus and the Double Door" is based on various translations of ancient Dionysian myths, and passages from some of the books mentioned above. The book quoted in regards to Plato's students, *A History of the World in 6 Glasses* by Tom Standage, is a great read and proved a valuable reference in the writing of several chapters.

And here's an excellent source that led me to other things, but I couldn't smoothly fit it into the bibliography: the Theoi Project's page on Icarius at http://www.theoi. com/Heros/Ikarios.html.

Anything to do with Dr. James Burke or Hangover Heaven—be it through interview, phone conversation, email or official website—was gathered and recorded during the winter of 2013.

FIRST INTERLUDE: A DRINK BEFORE THE WAR

Much of this section—in fact, much of this book—owes a great debt to Iain Gately's *Drink: A Cultural History of Alcohol*. It is the most enviable and the most comprehensive history of alcohol I have read, and it pointed out several historical avenues I wouldn't otherwise have taken. The Abraham Lincoln and David Lloyd George quotes came from Gately, while the Marco Polo quote was lifted from Holland's *Joy of Drinking*.

The Springsteen story is taken from listening to *Bruce Springsteen and the E Street Band Live, 1975–85,* as I've done a thousand times since getting it for Christmas when I was twelve years old. It is worth noting that Springsteen has never been much of a drinker and that, in telling this story, he never actually says that he was drinking, or had a hangover. That is just how I've always heard it.

PART TWO: WHAT HAPPENS ABOVE VEGAS

In "Going Down Drinking," the quotes from both Columella and Pliny the Elder were taken from Gately, while the historical synopses owe a debt to many of the publications already mentioned as well as some online encyclopedias.

I first learned of Levi Presley's Stratosphere suicide in a very roundabout way: through a number of magazine pieces about *The Lifespan of a Fact*, a book co-written by John D'Agata and Jim Fingal about a seven-year odyssey wherein Fingal attempted to properly fact-check an essay by D'Agata for *The Believer* that involved Presley's suicide.

SECOND INTERLUDE: PLENTY OF AVERSION: A VERSION OF PLINY

My research on Pliny came from historical biographies and encyclopedia entries that appear in the bibliography. But more important was the compiling of relevant entries in Pliny's own original encyclopedia, *Naturalis Historia*, with its dozens and dozens of volumes, in various translations. This, and so very much more, was generously and miraculously accomplished by my friend and researcher extraordinaire, Yannick Portebois. Her name should appear several times in these notes.

Although I rib them a bit in this section, I owe a debt of gratitude to Clement Freud, Keith Floyd and Andy Toper. As far as I could find, theirs are the only books on hang-overs and hangover history to date. And although all three are very slight, and mostly comical, I got a lot from them—including some points from which to start.

PART THREE: THE HAIR THAT WAGS THE DOG

As case in point on the previous note, the opening paragraph of "Man Drinks Dog" is due in part to Andy Toper, while the quote from Antiphanes is thanks to Keith Floyd.

The next few paragraphs, and other parts of the book, owe much to John Varriano's *Wine: A Cultural History*. I got a lot from this book about wine, art and death that I couldn't find elsewhere, including the stuff about the *Regimen Sanitatis Salernitanum*.

In much the same way, George Bishop (no relation) helped me a lot, with his very well-written, anachronistic and informative book *The Booze Reader*. Even if the only thing I'd got from it was the tonic made from crushed human skull, as found in *The London Distiller, 1667*, I'd owe him a great debt. But in fact, his book informed several parts of this one.

The bit after types and names of hairs of the dog—which were collected from pretty much everywhere—is thanks to Barbara Holland, particularly the quote from the National Institutes of Health.

The first acknowledgment to Adam Rogers's *Proof* comes at the end of "Man Drinks Dog." But I can't help but feel his excellent book about the science of alcohol also informed earlier sections—maybe even, in retrospect, "Welcome to Your Hangover." Rogers is a wonderful writer, and his book was incredibly helpful—as was he when I cold-called him anxiously one day, feeling like I didn't know anything. There are many sections after this that owe a debt to his expertise and generosity.

In regards to the medical forum "Asian Flush but Caucasian," it exists on the internet, but the posts that appear in this section are fictionalized.

While "'Let Moderation Reign!'" contains ideas accumulated from a number of the books referenced above, the quote from Arnold of Villanova is taken directly from Gately's *Drink*.

The quote from Avicenna, I got from Herbert M. Baus and his groundbreaking book *How to Wine Your Way to Good Health*. First introduced to me by my profoundly knowledgeable uncle, Mike Ross, Baus figured prominently in earlier drafts of this book. He is a wildly interesting figure—once a close advisor to Richard Nixon, then a red wine prophet both long after and long before it made sense to the masses—and his writings and sensibility still inform whole swaths of these pages.

As with Herbert M. Baus, Frank M. Paulsen also figured much larger in a previous draft of this book. While this overview and some samples still remain, it was hard to let so much of his truly awesome fieldwork go by the wayside. I encourage you to take a look, on the *Journal of American Folklore* website, under Paulsen and "A Hair of the Dog and Some Other Popular Hangover Cures from Popular Tradition." It is a remarkable read. The sections I've included here are with permission of the *Journal of American Folklore*, with thanks to Timothy Lloyd.

"A Reasonable Go Under the Circumstances" refers obliquely to several studies; their titles appear in the bibliography. Then it introduces the research of Dr. Richard Stephens, whom I later interviewed at length, in person. The interview appears at the beginning of Part Five, and his applicable studies, as well as his new book, are also in the bibliography.

I discovered the Herman Heise study regarding adrenaline in Bishop's *The Booze Reader*.

THIRD INTERLUDE: AND UP SHE RISES

The many nasty things in this section were culled from various places, but some of the most visceral are from Alice Morse Earl's 1896 book *Curious Punishments of Bygone Days*, Andrew Smith's *Drinking History: Fifteen Turning Points in the Making of American Beverages* and *A History of Alcoholism* by Jean-Charles Sournia, which was helpful throughout my research.

Of course, what brings this section together is the weird old work of Olaus Magnus the Goth. It was Toper's book that first tweaked me to this gothic magnus opus. As with Baus and Paulsen, there were once many more pages in this book devoted to the fixations of Olaus Magnus and his bizarre, judgmental tome. You should at least read the ninety-nine-word title, and also the chapter about honeybees who swarm drunken men—which, as with so much else, was found for me by Yannick Portebois.

PART FOUR: A MAD HATTER IN MIDDLE EARTH

The Glazebrook House Hotel in Devon, England, is where I stayed in the Jabberwocky Room, with the free, fully stocked bar. I bring this up because of the references in the text to this place, but also because it is one of the coolest hotels ever.

The official name of the museum in Boscastle is the Museum of Witchcraft and Magic. As I never did get to it, or follow up in any other way, I don't know the last name of Peter, who kindly responded to my email.

The Sydney Smith quote is from his essay "A Little Moral Advice."

In regards to "Wine and Cheese," Tim Spector was extremely helpful and gracious. His new book *The Diet Myth* ends with a chapter all about gut health that is excellent supplemental material to this section.

This is as good a place as any to thank *EnRoute* magazine and its editors, especially Susan Musgrave, for such interesting, on-point assignments, and support in this and other parts of the book.

About the two Samuels in "The Here and Hereafter": I believe I was first made aware of them by Iain Gately, then went down a rabbit hole of fire, brimstone and woodcuts—the various parts of which should be found in the bibliography. The point about Dante is from Sournia.

The best way to find what's been written about Nutton or Nutts (or written by them) is a simple Google search. In the end, they were both extremely generous sources, and I'm happy with how much of their interviews I was able to include.

In regards to the remedies referenced at the end of my conversation with Dr. Nutton, I have found historic references to anti-fume laurels and wreaths in too many places to mention. The idea persisted for centuries. I haven't, however, found anything to suggest that the mere presence of certain plants within the drinking environs could ever have an effect.

Then, admittedly, I've taken the following folkloric treatments mentioned by Freud and Toper at face value. Along with Robert Boyle's hemlock sock, they just didn't seem worth much scrutiny. And the reference to British voodoo witches and hangover effigies is, admittedly, based on pretty flimsy stuff, and is included mostly as a transition into Bilious the Oh God. He, of course, is real.

In "London Burning," the excellent thirteenth-century tourist quote is from Sournia, as is some of the information about the Gin Craze and industrialization's effects on alcohol consumption.

My list of of Britishisms for *drunk* is partially inspired by, and owes some credit to, Julian Baggini's 2003 article in the *Guardian*, "We Drink Therefore We Are."

In regards to the Hospital Club, I discovered the connection with John Harrison in Dava Sobel's book about him, *Longitude: The True Story of a Lone Genius Who Solved the Greatest Scientific Problem of His Time*. The information about the hotel's owners and Radiohead's studio I got from a London photographer with whom I was working on a magazine assignment.

FOURTH INTERLUDE: WEREWOLVES OF LONDON

William James's great quote about the "Yes" function is from his book *The Varieties of Religious Function*.

As alluded to in this section, it proved very difficult to find trustworthy sources in regards to the darker aspects of Chaney Sr. and Jr. It seems almost every substantive story written about either of them is in conflict with another—with very little documented information to back any of it up. It is worth pointing out that, despite his monumental contribution to Hollywood, the history of film and our culture in general, there has really never been a critically successful, comprehensive biography of Lon Chaney Jr.—which of course fits, creepily, with the life of a stillborn man who never got a grave. I have listed some books about him and his father in the bibliography, but in the end the source I found most helpful was a short bio of Chaney Jr. on the website HouseofHorrors.com.

And it should go without saying that this section's full potential relies on listening to Warren Zevon.

PART FIVE: TWELVE PINTS IN TWELVE PUBS

As mentioned earlier in these notes, Dr. Richard Stephens helped me out a lot. It was a pleasure to interview him, and his most applicable publications can be found in the bibliography.

This whole part of the book couldn't exist without the great generosity and help of two very specific camps. One is that of *The World's End,* including everyone at Big Talk Productions (especially Alex) and in the rights department of NBCUniversal—especially Roni Lubliner, who came through as if parting the seas for me—in regards to permission for so much use of such a great film. The other camp is that of Dart and Son. Without my great friends and allies, Jonathan and Thomas Dart, this part of the book, which I may have enjoyed more than any other, never would have happened. Their knowledge, expertise and bravery were invaluable.

The many articles quoted and referred to in this part can be found by author's name in the bibliography. But how I found many of them, along with sources in other parts of the book, is thanks to one particularly impressive publication: *Smashed!: The Many Meanings of Intoxication and Drunkenness* by Peter Kelly, Jenny Advocat, Lyn Harrison and Christopher Hickey.

The Andrew Anthony quotes are from his article in the *Guardian* on October 5, 2004.

As with Peter of the witchcraft museum, I never did discover the last name of the bad witch.

FIFTH INTERLUDE: THE WITHNAIL AWARDS: A PRESS RELEASE

In regards to "Best Hungover Dialogue," the permission for lines from both *Anchorman* and *True Grit* were generously provided by Paramount Pictures, with special thanks to Larry McCallister—for the second time in these citations. Permission to use the line from *My Favorite Year* was granted by Warner Bros. Entertainment Inc., with special thanks to Shannon Fifer. The usage of dialogue from *Die Hard: With a Vengeance* was provided by 20th Century Fox, with thanks to Andy Bandit.

PART SIX: THE HUNGOVER GAMES

The brilliant bit of dialogue from Danny Boyle's epic adaptation of Irvine Welsh's *Trainspotting* was generously provided by Figment Films, with great thanks to Joanne Smith.

There are many sources, and also videotape, for Stan Bowles's disastrous day on *The Superstars*, but if you want to get it from the horse's mouth, Bowles ran it all down for the *Guardian* in a first-person recap on May 20, 2009.

So taken was I by Shaan Joshi's narrative description of Max McGee's first-ever Super Bowl hungover heroics that I found myself mimicking his style in an anecdote near the end of this book. See if you can spot where.

The bulk of my research at Speyside and interviews with Brian Kinsman were thanks to *Sharp's Book for Men*, particularly Greg Hudson; Jesson and Company, particularly Trevor Walsh; and Beth Havers of Glenfiddich Canada. Beyond that, it was really Malt Master Kinsman who helped me learn what I was looking for in the Highlands, and then farther on. His expertise regarding distillation was supplemented with passages from Bishop; Rogers; *Alcohol: Its History, Pharmacology and Treatment* by Mark Edmund Rose and Cheryl J. Cherpitel, as well as *Chemical Additives in Booze* by Michael Jacobson and Joel Anderson.

I found the Scottish Board of Health poster—or, at least, a clear photo of it—in the pages of Ian Middleton's impressive self-published paper "A Short History of the Temperance Movement in the Hillfoots." This is also where I found the information about licensing acts, including the one about "bona fide travellers" on Sundays.

I learned a heckuva lot about eighteenth- and nineteenth-century temperance propaganda created specifically for children—and got to browse through original publications of Cruikshank and others—thanks to the librarians of the Osborne Collection at the Lillian H. Smith Library in Toronto.

Buckie became mainstream news in Scotland in 2013, thanks to articles like the one by Auslan Cramb in the *Telegraph*, which ran on December 27.

The testimonials in "A Helluva Way to Wake Up" were solicited by me through social media.

SIXTH INTERLUDE: A ROOTS OF REMEDY ROUNDUP

This being a roundup, it is the synthesis of too many sources to reasonably track. But the quote in regard to being "in need of a herring," as well as Aristotle's rhyme about cabbage, are taken from Freud. The quote can be found in Gillian Riley's book *Food in Art: From Prehistory to the Renaissance*, while the stuff about the "herring game" is from Gately. The full text of NoHo's patent is available online, as are the results of its clinical tests.

PART SEVEN: THE FUTURE'S SO BRIGHT

While all of my brave test subjects at the St. Paddy's Day party and otherwise gave their permission to be quoted, one or two of them asked that their names be changed for publication.

My somewhat unconventional ideas about red wine, pesticides and migraines have been bolstered if not refined during dozens of conversations with organic winemakers and chemists over the years. But what precipitated all that was a chat I had with Bruce Ewart of L'Acadie Vineyards in Wolfville, Nova Scotia.

"Welcome to the Monkey House" was cobbled together using the following articles and more—most all of it discovered by the tireless Yannick Portebois: "Man Loses License After Drink-Driving in Toy Barbie car," *Telegraph*, April 19, 2010; "Man Who Legally Changes His Name to Bacon Double Cheeseburger: 'I've got No Regrets at All,'" *The Comeback*, February 24, 2016; "Blame It on the Alcohol? Maybe Not, Study Suggests," NBC News, September 22, 2011; "Ten Hangover Nightmares: Those Who Lived to Regret the Night Before," *Telegraph*, April 23, 2014.

A select few of Joris Verster's publications can be found in the bibliography, while many more can be found online.

SEVENTH INTERLUDE: KILLER PARTIES

Among other publications listed in the bibliography, this section most owes a debt to *The Poisoner's Handbook* by Deborah Blum, as well as her article "The Chemist's War" in *Slate*, February 19, 2010.

PART EIGHT: THE TIGER ON THE ROOF

Most of the access and information, as well the permission for photo use in this part is due to the generosity of acclaimed international photographer David Yarrow, and also that of Cason Thorsby.

The information on Detroit during Prohibition came from a number of sources, but among the most helpful was *Detroit Underground History: Prohibition and the Purples* by Shannon Saksewski.

I did a lot of baseball reading for "The Middle Ball," including Robert Creamer's *Babe: The Legend Comes to Life*, Allen Barra's *Mickey and Willie: Mantle and Mays, The Parallel Lives of Baseball's Golden Age*, Mickey Mantle's own *Sports Illustrated* confessional "Time in a Bottle" from April 18, 1994, David Wells's autobiography *Perfect, I'm Not*, and parts of Bert Randolph Sugar's book *The Great Baseball Players: From McGraw to Mantle*.

Along with the sources and authors referenced in "The Taste of Freedom," it owes a debt—as do some later sections—to Olivia Laing. Her book *The Trip to Echo Springs* is one of the best ever about male writers and drinking, but it was her essay in the *Guardian*, "Every Hour a Glass of Wine," about great female writers in this regard, that inspired and helped many of the ideas here.

EIGHTH INTERLUDE: I WOKE UP THIS MORNING

The first paragraph of this section is a sort of early-morning mash-up of Bessie Smith, Bill Withers, Janis Joplin, Saves the Day, Shania Twain, Bruce Springsteen, Nazareth, Peter Frampton, the Byrds, the Descendents, Phil Collins, Rufus Wainwright, Sting, Howlin' Wolf, Bob Dylan and John Lennon, among others.

Much of the phrasing of "the quintessential one" and its place in popular music can be attributed to B.B. King.

The lyrics to the Chicago song "An Hour in the Shower—A Hard Risin' Morning Without Breakfast" have been reprinted here with the extremely generous permission of Terry Kath's daughter Michelle Kath. I will always feel honoured and grateful for that.

Part Nine: Beyond the Volcanoes

Although the aches and pains might suggest otherwise, I did eventually enjoy my stay at Rogner Bad Blumau very much. It is a remarkable place. The information provided by various pamphlets and signs at Rogner Bad Blumau, and quoted here, was also supplemented through interviews with staff and the charming, helpful onsite director and manager, Lucy Seteram.

Much of my itinerary throughout the trip through Austria—including habitation and interviews—was organized by the estimable Ms. Rita Gily, who seems to know everyone in the Alps. She also helped with some of the research.

While the information in "A Whole Lot of Bull" came from numerous articles, the quote from Ted Farnsworth, CEO of Purple, can be found in a MarketWatch story, "Liquored up and Lively in Las Vegas," from February 8, 2008.

The experience and information, albeit vague, surrounding the Kräuter-Heubad, or herbal hay bath, comes from my visit and interviews at the Almcorf Seinerzeit Resort, where my contact was Ms. Betina Welter.

"The Katers of the World" owes not only an obvious debt to the artist Dennis Schuster for his awesome images and Martin Breuer for his ad-world vision, but also others in the offices of Thomapyrin and Havas Worldwide, particularly Jochen Schwarz when it came to securing permissions. Thanks also to Philipp Geissler for his artful translations.

These last sections of this part also owe a debt of thanks, and some help with research, to the various entities of Hangover Hostel, the Stoke Travel Company and also my great friend Tracy Fladl and her wonderful family.

Ninth Interlude: Aspirin of Sorrow

This section is the result of reading not only hundreds of novels with hangovers in them, but dozens of scenes involving Aspirin. As with much of this book, Yannick Portebois was invaluable, and also did much of the reading. What appears here is the very tip of the point of the iceberg.

Part Ten: When Lizards Drink from Your Eyes

The narrative in this part of the book represents a personal recollection. Bits have been left out and some names changed to protect people's privacy.

In "Elpenor on the Roof," while the stories about Odysseus and Elpenor are taken from readings of Homer's *Odyssey*, I first encountered the idea of Elpenor's syndrome in Adam Rogers's *Proof.*

The combination of awful things referred to in "Your Worst Hangover Ever" are the result of years and years of searching out hangover stories. The ER case studies, however, were very specifically brought to my attention—along with about a hundred others—by my friend, fellow writer and astute researcher David Lightfoot.

Although I never do quote from it (I thought if I did, I'd never stop), this book does owe a great debt to the movie *The Hangover*. I enjoyed watching and rewatching it, and also reading about it on so many internet sites. Meanwhile, the specs on the depressing story about Siobhan Watson's wedding day—also discovered for me by Yannick Portebois—can be found in the bibliography.

Tenth Interlude: The Hangover Writer

The opening quote by Nietzsche is from *Twilights of the Idols, or How to Philosophize with a Hammer*, while the quote by Horace is . . . well, from Horace.

The Roald Dahl quote is from "Tales of Childhood," and the line from *Arthur* is reproduced with the generous permission of Warner Bros. Entertainment Inc., with thanks to Shannon Fifer.

The Malcolm Lowry quote is from his unfinished (or lost) novel, *Ballast to the White Sea*, a manuscript that was only recently discovered and published in 2014.

The quote from Zachary Leader, who also assisted with some research questions, is from Amis's official biography *The Life of Kingsley Amis*. The other biographies of Amis, as well as autobiographies of Clement Freud and Keith Floyd that I read for both this section and personal interest, can be found in the bibliography.

Part Eleven: After the Flood

I owe a great debt of thanks—for the idea to finish this book in New Orleans, organizing our trip and helping with the research, as well as just helping me survive to the end of this book—to the lovely, miraculous Angela McDonald.

Marita Jaeger of Boutique de Vampyre was very helpful. She now goes by Marita Woywod Crandle and has recently opened up a "magical speakeasy" that serves potions. It's on Bourbon Street, and you need a password to get in. You might get some clues in this regard at PotionsLounge.com.

Much of the historical research in this last part started at the New Orleans Pharmacy Museum. It is a wondrous place full of curios, oddities, potions, poisons, questionable advertising and obscure facts you'd be hard-pressed to find anywhere else. The passages pertaining to patent medicines, nerve tonics, absinthe, snake oil and Dudley J. Leblanc owe a debt to some corner of this large, creaky house.

The names of many brave people who let me test different hangover remedies on them over the years can be found in the acknowledgments. These tests and studies took place at house parties, poker games, bars, bar conventions, wine tours and weddings over a four-year period.

My interviews with Dr. Mignonne Mary of the Remedy Room were crucial in the final stages of my research and in the beginning of refining my own "cure." Her coworker Alison Frankel was also a great help.

Liz Williams, president of the American Cocktail Museum, was a helpful and gracious host. As were the many, many bartenders I drank and spoke with throughout the French Quarter.

And of course, thanks are due to Dr. Nutt and Miss Mambo Marie.

FOR THE LOVE OF HANGOVERS: A KIND OF CONCLUSION

Most of the general ideas in this final section are merely a synthesis of years of research as referenced in the book itself, the above notes and the bibliography.

My main sources for the stories referenced in this conclusion are as follows: "This Company Is Offering Staff Free 'Hangover Days,'" Rosie Fitzmaurice, *Business Insider UK*, August 25, 2017; "Hungover Customer Brings Heaps of Business to Struggling Alberta Fish and Chips Shop." Danielle Nerman, CBC News, August 26, 2016; "Throwing It All away," Andrew Mayer, CBC.ca, April, 2017; "Toronto Crane Girl Pleads Guilty: 'I Thought It Would Make Me Feel More Alive,'" Betsy Powell, *Toronto Star*, January 10, 2018; "Stolen Statue Returned to Kensington Market Church With an Apology," Michelle Lepage, *Toronto Star*, December 5, 2013.

The last few pages, of course, owe much to the silver-haired lady.

BIBLIOGRAPHY

Abel, Ernest L. *Intoxication in Mythology: A Worldwide Dictionary of Gods, Rites, Intoxicants and Places.* Jefferson, NC: McFarland, 2006.

ABMRF: The Foundation for Alcohol Research. *Moving Forward* (2014 Annual Report).

Abram, Christopher. *Myths of the Pagan North: The Gods of the Norsemen.* London: Continuum, 2011.

Amis, Kingsley. *Everyday Drinking.* New York: Bloomsbury, 2008.

———. *Lucky Jim.* New York: Doubleday, 1954.

Anthony, Andrew. "Will Bladdered Britain Ever Sober Up?" *Guardian* (London), October 5, 2004.

Arumugam, Nadia. "Wine Scams: The Ultimate Hall of Fame." *Forbes*, January 8, 2013.

Association against the Prohibition Amendment. *Canada Liquor Crossing the Border.* Washington, DC: Association against the Prohibition Amendment, 1929.

Ayto, John. *The Diner's Dictionary: Word Origins of Food and Drink*, 2nd ed. Oxford: Oxford University Press, 2012. Published online 2013. http://www.oxfordreference.com/view/10.1093/acref/9780199640249.001.0001/acref-9780199640249.

Barnard, Mary. "The God in the Flowerpot." *American Scholar*, Autumn 1963.

Barra, Allen. *Mickey and Willie: Mantle and Mays, The Parallel Lives of Baseball's Golden Age.* New York: Crown Archetype, 2013.

Baus, Herbert M. *How to Wine Your Way to Good Health.* New York: Mason and Lipscomb, 1973.

Bishop, George. *The Booze Reader: A Soggy Saga of Man in His Cups.* Los Angeles: Sherbourne Press, 1965.

Blake, Michael F. *Lon Chaney: The Man Behind the Thousand Faces.* New York: Vestal Press, 1993.

Blakemore, Colin, and Sheila Jennett. *The Oxford Companion to the Body.* Oxford: Oxford University Press, 2001.

Blocker, Jack S., David M. Fahey and Ian R. Tyrrell, eds. *Alcohol and Temperance in Modern History*. Santa Barbara, CA: ABC-CLIO, 2003.

Blum, Deborah. "The Chemist's War." *Slate*, February 19, 2010. http://www.slate.com/articles/health_and_science/medical_examiner/2010/02/the_chemists_war.html.

———. *The Poisoner's Handbook*. New York: Penguin, 2010.

Boyle, Robert. *Medicinal Experiments*. London: Samuel Smith and B. Walford, 1698.

Braun, Stephen. *Buzz: The Science and Lore of Alcohol and Caffeine*. New York: Oxford University Press, 1996.

Bukowski, Charles. "Everything." In *The Roominghouse Madrigals: Early Selected Poems 1946–1966*. New York: Ecco, 2002.

———. *Factotum*. Santa Barbara, CA: Black Sparrow Press, 1982.

Burchill, Julie. "The Pleasure Principle." *Guardian* (London), December 1, 2001.

Burns, Eric. *The Spirits of America: A Social History of Alcohol*. Philadelphia: Temple University Press, 2004.

Burton, Kristen D. "Blurred Forms: An Unsteady History of Drunkenness." *Appendix* 2, no. 4 (October 2014). http://theappendix.net/issues/2014/10/blurred-forms-an-unsteady-history-of-drunkenness.

Carey, Sorcha. *Pliny's Catalogue of Culture: Art and Empire in the Natural History*. Oxford: Oxford University Press, 2003.

Cato, Marcus Porcius. *De Agricultura*. Cambridge, MA: Harvard University Press, 1934.

Chapman, Carolynn. "The Queen Mother Averaged More than 70 Drinks a Week." *Whiskey Goldmine*, February 9, 2011.

Clark, Lindsay D. "Confrontation with Death Illuminates Death's Mystery in the *Odyssey*." *Inquiries Journal 1*, no. 11 (2009). www.inquiriesjournal.com/articles/71/confrontation-with-death-illuminates-deaths-mystery-in-the-odyssey.

Creamer, Robert. *Babe: The Legend Comes to Life*. New York: Simon and Schuster, 1974.

Crewe, Daniel. "'One of Nature's Liberals': A Biography of Clement Freud." *Journal of Liberal History* 43 (Summer 2004): 15–18.

Crofton, Ian. *A Dictionary of Scottish Phrase and Fable*. Edinburgh: Birlinn, 2012.

Crosariol, Beppi. "Should You Be Worried about Pesticides in Wine?" *Globe and Mail* (Toronto), August 31, 2011.

Crozier, Frank P. *A Brass Hat in No Man's Land*. New York: J. Cape and H. Smith, 1930.

Dahl, Roald. *Tales of Childhood*. London: Penguin, 1984.

Dalby, Andrew. *Bacchus: A Biography*. London: British Museum Press, 2003.

Davidson, Alan. *The Oxford Companion to Food*. Oxford: Oxford University Press, 2014.

Davidson, James. *The Greeks and Greek Love: A Bold New Exploration of the Ancient World*. New York: Random House, 2007.

Davis, Kenneth C. *Don't Know Much About Mythology*. New York: HarperCollins, 2005.

de Haan, Lydia, Hein de Haan, Job van der Palen and Joris C. Verster. "The Effects of Consuming Alcohol Mixed with Energy Drinks (AMED) Versus Consuming Alcohol Only on Overall Alcohol Consumption and Alcohol-Related Negative Consequences." *International Journal of General Medicine* 5 (2012): 953–60.

Dent, Susie, ed. *Brewer's Dictionary of Phrase and Fable*, 19th ed. London: Chambers Harrap, 2012. Published online 2013. http://www.oxfordreference. com/view/10.1093/acref/9780199990009.001.0001/acref-9780199990009.

Devitt, Brian M., Joseph F. Baker, Motaz Ahmed, David Menzies and Keith A. Synnott. "Saturday Night Palsy or Sunday Morning Hangover? A Case Report of Hangover-Induced Crush Syndrome." *Archives of Orthopaedic and Trauma Surgery* 131, no. 1 (January 2011): 39–43.

Dodd, C.E. "Lectures at the Incorporated Law Society—Notes of Lectures by C.E. Dodd, esq.—On the Constitution of Contracts.—Assent.— Construction [regarding contracts signed while drunk]." *Legal Observer, Or, Journal of Jurisprudence* 12 (July 1836).

Down, Alex. "Austrian Wine: From Ruin to Riches." *Drinks Business,* February 13, 2014. https://www.thedrinksbusiness.com/2014/02/austrian-wine-from-ruin-to-riches/.

Earl, Alice Morse. *Curious Punishments of Bygone Days*. Chicago: H.S. Stone, 1896; Bedford, MA: Applewood Books, 1995.

Edwards, Griffith. *Alcohol: The World's Favorite Drug*. New York: Thomas Dunne, 2000.

Ekirch, Robert. *At Day's Close: Night in Times Past*. New York: Norton, 2005.

Elias, Megan J. *Food in the United States, 1890–1945*. Westport, CT: Greenwood, 2009.

Ernst, Edzard. "Detox: Flushing Out Poison or Absorbing Dangerous Claptrap?" *Guardian* (London), August 29, 2011.

Floyd, Keith. *Floyd on Hangovers*. London: Penguin, 1992.

———. *Stirred but Not Shaken: The Autobiography*. London: Sidgwick and Jackson, 2009.

Frankenberg, Frances R. "It's Not Easy Being Emperor." *Current Psychiatry* 5, no. 5 (May 2006): 73–80.

Franks, General Tommy. *American Soldier*. New York: HarperCollins, 2003.

Freud, Clement. *Clement Freud's Book of Hangovers.* London: Sheldon Press, 1981.

———. *Freud Ego*. London: BBC Worldwide, 2001.

Fuller, Robert C. *Religion and Wine: A Cultural History of Wine Drinking in the United States*. Knoxville, TN: University of Tennessee Press, 1996.

Gagarin, Michael, ed. *The Oxford Encyclopedia of Ancient Greece and Rome*. Oxford: Oxford University Press, 2012.

Gately, Iain. *Drink: A Cultural History of Alcohol*. New York: Gotham Books, 2008.

Gauquelin, Blaise. "Les buveurs de schnaps n'ont qu'à bien se tenir." *Libération*, September 5, 2014.

Glyde, Tania. "The Longest Hangover in My 23 Years as an Alcoholic." *Independent*, January 18, 2008.

Goodwin, Donald W. "Alcohol as Muse." *American Journal of Psychotherapy* 46, no. 3 (July 1992): 422–33.

Gopnik, Adam. "Writers and Rum." *New Yorker*, January 9, 2014.

Graber, Cynthia. "Snake Oil Salesmen Were on to Something." *Scientific American*, November 1, 2007. https://www.scientificamerican.com/article/snake-oil-sales-men-knew-something/.

Green, Harriet. "Gruel to Be Kind: A Hardcore Detox Break in Austria." *Guardian* (London), January 12, 2013.

Green, Jonathon. *Green's Dictionary of Slang*. London: Chambers Harrap, 2010. Published online 2011. http://www.oxfordreference.com/view/10.1093/acref/9780199829941.001.0001/acref-9780199829941.

Gutzke, David W. *Women Drinking Out in Britain Since the Early Twentieth Century*. Manchester: Manchester University Press, 2014.

Halberstadt, Hans. *War Stories of the Green Berets*. Saint Paul, MN: Zenith Press, 2004.

Hannaford, Alex. "Boozed and Battered." *Guardian* (London), January 20, 2004.

Harbeck, James. "Hangover." *Sesquiotic*, January 1, 2011. https://sesquiotic. wordpress.com/2011/01/01/hangover/.

Hatfield, Gabrielle. *Encyclopedia of Folk Medicine: Old World and New World Traditions*. Santa Barbara, CA: ABC-CLIO, 2004.

Haucap, Justus, Annika Herr and Björn Frank. "In Vino Veritas: Theory and Evidence on Social Drinking" (DICE Discussion Paper No. 37). Düsseldorf, Germany: Düsseldorf Institute for Competition Economics, 2011.

Hemingway, Ernest. *A Farewell to Arms*. New York: Scribner, 1929.

———. *A Moveable Feast*. New York: Scribner, 1964.

Henley, Jon. "Bonjour Binge Drinking." *Guardian* (London), August 27, 2008.

Holland, Barbara. *The Joy of Drinking*. New York: Bloomsbury, 2007.

Holmes, Richard, Charles Singleton and Spencer Jones, eds. *The Oxford Companion to Military History*. Oxford: Oxford University Press, 2001.

Hornblower, Simon, and Tony Spawforth, eds. *Who's Who in the Classical World*. Oxford: Oxford University Press, 2003.

Hough, Andrew. "Keith Floyd Dies: The Outspoken Television Chef Has Died after a Heart Attack." *Telegraph* (London), September 15, 2009.

Huzar, Eleanor. "The Literary Efforts of Mark Antony." In *Aufstieg und Niedergang der römischen Welt*, edited by Hildegard Temporini and Wolfgang Haase, 639–57. Berlin: Walter de Gruyter, 1982.

Irvine, Dean. "When Massages Go Bad." *CNN Project Life*, June 13, 2007. http://www.cnn.com/2007/HEALTH/05/22/pl.massagegobad/index.html.

Ísleifsson, Sumarliði R., and Daniel Chartier, eds. *Iceland and Images of the North*. Montreal: Presses de l'Université du Québec, 2011.

J.F., "A New Letter, to All Drunkards Whoremongers, Thieves, Disobedience to Parents, Swearers, Lyers, &c.: Containing a Serious and Earnest Exhortation that They Would Forsake Their Evil Ways." London: F. Bradford. 1695.

Jacobson, Michael F., and Joel Anderson. *Chemical Additives in Booze.* Washington, DC: Center for Science in the Public Interest, 1972.

James, William. *The Varieties of Religious Experience: A Study in Human Nature.* New York: Longmans, Green, 1902; n.p.: Renaissance Classics Press, 2012.

Jivanda, Tomas. "A Bottle of Wine a Day Is Not Bad for You and Abstaining Is Worse than Drinking, Scientist Claims." *Independent* (London), April 19, 2014.

Jodorowsky, Alejandro. *Psychomagic: The Transformative Power of Shamanic Psychotherapy.* New York: Simon and Schuster, 2010.

Jones, Stephen. *The Illustrated Werewolf Movie Guide.* London: Titan, 1996.

Joshi, Shaan. "Max McGee Goes Out Drinking: The Story of a Super Bowl Legend." *Prague Revue,* January 31, 2013.

Karibo, Holly M. *Sin City North: Sex, Drugs and Citizenship in the Detroit–Windsor Borderland.* Chapel Hill, NC: University of North Carolina Press, 2015.

Kelly, Peter, Jenny Advocat, Lyn Harrison and Christopher Hickey. *Smashed! The Many Meanings of Intoxication and Drunkenness.* Clayton, Australia: Monash University Publishing, 2011.

Kennedy, William. *Ironweed.* New York: Viking, 1983.

Kerouac, Jack. *Big Sur.* New York: Farrar, Straus and Cudahy, 1962; New York: Penguin, 1992.

Laing, Olivia. "Every Hour a Glass of Wine: The Female Writers Who Drank." *Guardian* (London), June 13, 2014.

———. *The Trip to Echo Spring: On Writers and Drinking.* New York: Picador, 2013.

"LeBlanc Medicine Co., Docket No. 6390," in Federal Trade Commission, *Annual Report for the Fiscal Year Ended June 30, 1955:* 41–42.

Lecky, William Edward Hartpole. *A History of England in the 18th Century,* vol. 1. London: Longman, Green, 1878.

Lesieur, O., V. Verrier, B. Lequeux, M. Lempereur and E. Picquenot. "Retained Knife Blade: An Unusual Cause for Headache Following Massive Alcohol Intake." *Emergency Medicine Journal* 23, no. 2 (February 2006): e13.

Liberman, Sherri, ed. *American Food by the Decades.* Westport, CT: Greenwood, 2011.

Lindow, John. *Norse Mythology: A Guide to Gods, Heroes, Rituals and Beliefs.* Oxford: Oxford University Press, 2002.

London, Jack. *John Barleycorn.* New York: Century, 1913; New York: Modern Library, 2001.

Lowry, Malcolm. *In Ballast to the White Sea.* Ottawa: University of Ottawa Press, 2014.

———. *Under the Volcano.* New York: Reynal and Hitchcock, 1947; New York: Perennial Classics, 2000.

Magnus, Olaus. *Description of the Northern Peoples.* [In Latin.] Translated by Peter Fisher and Humphrey Higgens. Edited by Peter Foote. London: Hakluyt Society, 1996.

Mankiller, Wilma, Gwendolyn Mink, Marysa Navarro, Barbara Smith and Gloria Steinem, eds. *The Reader's Companion to U.S. Women's History.* Boston: Houghton Mifflin, 1998. See esp. "Alcoholism" (p. 24) and "Prohibition" (p. 479).

Mantle, Mickey. "Time in a Bottle." *Sports Illustrated,* April 18, 1994.

Marshall, Sarah. "Don't Even Brush Your Teeth: 91 Hangover Cures from 1961." *Awl,* July 11, 2012. https://medium.com/the-awl/dont-even-brush-your-teeth-91-hangover-cures-from-1961-88353fe97fcc.

Martelle, Scott. *Detroit: A Biography.* Chicago: Chicago Review Press, 2012.

Martinez-Carter, Karina. "Fernet: The Best Liquor You're (Still) Not Yet Drinking." *Atlantic,* December 30, 2011.

Mason, Philip P. *Rum Running and the Roaring Twenties: Prohibition on the Michigan–Ontario Waterway.* Detroit: Wayne State University Press, 1995.

Middleton, Ian. "A Short History of the Temperance Movement in the Hillfoots." Ochils Landscape Partnership. http://ochils.org.uk/sites/default/files/oral-histories/docs/temperance-essay.pdf.

Nash, Thomas. *Pierce Penilesse: His Supplication to the Devil. Describing the Overspreading of Vice, and Suppression of Virtue. Pleasantly Interlaced with Variable Delights, and Pathetically Intermixed with Conceited Reproofs.* London: Richard Jones, 1592.

Nelson Evening Mail (Nelson, New Zealand). "Curing Drunkards by Bee-Stings." July 11, 1914.

Nietzsche, Friedrich. *Twilights of the Idols, or How to Philosophize with a Hammer.* [In German.] Oxford: Oxford University Press, 1998.

Nordrum, Amy. "The Caffeine-Alcohol Effect." *Atlantic,* November 7, 2014.

Norrie, Philip. "Wine and Health through the Ages with Special Reference to Australia." PhD diss., University of Western Sydney School of Social Ecology and Lifelong Learning, 2005.

Nutt, David. "Alcohol Alternatives—A Goal for Psychopharmacology?" *Journal of Psychopharmacology* 20, no. 3 (2006): 318–20. And other applicable papers.

Nutton, Vivian. *Ancient Medicine.* New York: Routledge, 2004. And several applicable papers.

O'Brien, John. *Leaving Las Vegas.* New York: Grove Press, 1990.

Orchard, Andy. *Dictionary of Norse Myth and Legend.* London: Cassell, 1996.

Osborne, Lawrence. *The Wet and the Dry: A Drinker's Journey.* New York: Crown, 2013.

Ovid. *Metamorphoses.* Translated by Rolfe Humphries. Bloomington, IN: Indiana University Press, 1955.

Palmer, Brian. "Does Alcohol Improve Your Writing?" *Slate*, December 16, 2011. http://www.slate.com/articles/news_and_politics/explainer/2011/12/christoper_hitchens_claimed_drinking_helped_his_writing_is_that_true_.html.

Paulsen, Frank M. "A Hair of the Dog and Some Other Hangover Cures from Popular Tradition." *Journal of American Folklore* 74, no. 292 (April–June 1961): 152–68.

Peck, Garrett. *The Prohibition Hangover: Alcohol in America from Demon Rum to Cult Cabernet*. New Brunswick, NJ: Rutgers University Press, 2009.

Perry, Lacy. "How Hangovers Work." HowStuffWorks.com, October 12. 2004. https://health.howstuffworks.com/wellness/drugs-alcohol/hangover.htm.

Pittler, Max H., Joris C. Verster and Edzard Ernst. "Interventions for Preventing or Treating Alcohol Hangover: Systematic Review of Randomized Control Trials." *British Medical Journal* 331, no. 7531 (December 24–31, 2005): 1515–17.

Plack, Noelle. "Drink and Rebelling: Wine, Taxes, and Popular Agency in Revolutionary Paris, 1789–1791." *French Historical Studies* 39, no. 3 (August 2016): 599–622.

Pliny. *Natural History*. [In Latin.] Translated by H. Rackham (vols. 1–5, 9), W.H.S. Jones (vols. 6–8) and E.E. Eichholz (vol. 10). Cambridge, MA: Harvard University Press, 1938. Reprinted 1967.

Pratchett, Terry. *The Hogfather*. New York: HarperPrism, 1996.

Rae, Simon, ed. *The Faber Book of Drink, Drinkers, and Drinking*. London: Faber and Faber, 1991.

Ramani, Sandra. "Top 10 Booze-Infused Spa Treatments." *Fodor's Travel*, April 24, 2013. https://www.fodors.com/news/top-10-boozy-spa-treatments-6574.

Rhosenow, Damaris J., Jonathan Howland, Sara J. Minsky, Jacey Greece, Alissa Almeida and Timothy A. Roehrs. "The Acute Hangover Scale: A New Measure of Immediate Hangover Symptoms." *Addictive Behaviors* 32, no. 6 (June 2007): 1314–20.

Reid, Stuart J. *A Sketch of the Life and Times of the Rev. Sydney Smith*. London: Sampson Low, Marston, Searle, and Rivington, 1884.

Rich, Frank Kelly. *Modern Drunkard* magazine. https://drunkard.com/

Riley, Gillian. *Food in Art: From Prehistory to the Renaissance*. London: Reaktion Books, 2015.

Robertson, Brandon M., Thomas M. Piasecki, Wendy S. Slutske, Phillip K. Wood, Kenneth J. Sher, Saul Shiffman and Andrew C. Heath. "Validity of the Hangover Symptoms Scale: Evidence from an Electronic Diary Study." *Alcoholism Clinical & Experimental Research* 36, no. 1 (January 2012): 171–77.

Rogers, Adam. *Proof: The Science of Booze*. Boston: Houghton Mifflin Harcourt, 2015.

Rose, Mark Edmund, and Cheryl J. Cherpitel. *Alcohol: Its History, Pharmacology and Treatment*. Center City, MN: Hazelden, 2011.

Saksewski, Shannon. "Detroit Underground History: Prohibition and the Purples." *Awesome Mitten*, February 25, 2014. https://www.awesomemitten.com/detroit-underground-history/.

Schneider, Stephen. *Iced: The Story of Organized Crime in Canada.* Mississauga, ON: Wiley, 2009.

Schoenstein, Ralph, ed. *The Booze Book: The Joy of Drink.* Chicago: Playboy Press, 1974.

Scott, Kenneth. "Octavian's Propaganda and Antony's *De Sua Ebrietate.*" *Classical Philology* 24, no. 2 (April 1929): 133–41.

Shakar, Alex. *Luminarium.* New York: Soho Press, 2011.

Shubin, Neil. *Your Inner Fish: A Journey into the 3.5-Billion-Year History of the Human Body.* New York: Vintage Books, 2009.

Sinclair, Andrew. *Prohibition: The Era of Excess.* Boston: Little, Brown, 1962.

Smith, Andrew. *Drinking History: Fifteen Turning Points in the Making of American Beverages.* New York, Columbia University Press, 2014.

Smith, William. *Dictionary of Greek and Roman Biography and Mythology.* London: Taylor, Walton, and Maberly, 1870.

Sobel, Dava. *Longitude: The True Story of a Lone Genius Who Solved the Greatest Scientific Problem of His Time.* New York: Bloomsbury, 2007. First published in 1995 by Walker (New York).

Sournia, Jean-Charles. *A History of Alcoholism.* [In French.] Translated by Nick Hindley and Gareth Stanton. Oxford: Basil Blackwell, 1990.

Spector, Tim. *The Diet Myth: The Real Science Behind What We Eat.* London: Weidenfeld & Nicolson, 2015.

———. "Why Is My Hangover So Bad?" *Guardian* (London), June 21, 2015. https://www.theguardian.com/lifeandstyle/2015/jun/21/why-is-my-hangover-so-bad.

Standage, Tom. *A History of the World in 6 Glasses.* New York: Walker, 2006.

Stephens, Richard. *Black Sheep: The Hidden Benefits of Being Bad.* London: Hodder and Stoughton, 2015. Also many applicable papers.

Stevenson, Robert Louis. *The Strange Case of Dr. Jekyll and Mr Hyde and Other Tales of Terror.* London: Penguin Classics, 2002.

Stöckl, Albert. "Australian Wine: Developments after the Wine Scandal of 1985 and Its Current Situation." Paper presented at the 3rd International Wine Business Research Conference, Montpellier, France, July 6–8, 2006.

Stone, Jon. "Beer Day Britain: How the Magna Carta Created the Humble Pint." *Independent* (London), June 15, 2015. https://www.independent.co.uk/news/uk/home-news/the-magna-cartas-role-in-creating-the-humble-pint-of-beer-10320844.html.

Sugar, Bert Randolph. *The Great Baseball Players: From McGraw to Mantle.* Mineola, NY: Dover Publications, 1997.

Swift, Robert, and Dena Davidson. "Alcohol Hangover: Mechanisms and Mediators." *Alcohol Health & Research World* 22, no. 1 (1998): 54–60.

Tagliabue, John. "Scandal over Poisoned Wine Embitters Village in Austria." *New York Times*, August 2, 1985.

Thomas, Caitlin. *Double Drink Story: My Life with Dylan Thomas*. London: Virago Press, 1998.

———. *Leftover Life to Kill*. London: Putnam, 1957.

Thompson, Derek. "The Economic Cost of Hangovers." *Atlantic*, July 5, 2013. https://www.theatlantic.com/business/archive/2013/07/the-economic-cost-of-hangovers/277546/.

Toper, Andy. *The Wrath of Grapes, or The Hangover Companion*. London: Souvenir Press, 1996.

United Kingdom. *Hansard Parliamentary Debates*, 3d series, vol. 353 (1891), cols. 1701–1707.

Vallely, Paul. "2,000 Years of Binge Drinking." *Independent* (London), November 19, 2005. https://www.independent.co.uk/news/uk/this-britain/2000-years-of-binge-drinking-516009.html.

Valliant, Melissa. "Do Juice Cleanses Work? 10 Truths about the Fad." *Huffington Post*, March 22, 2012. http://www.huffingtonpost.ca/2012/03/22/do-juice-cleanses-work_n_1372305.html.

Varriano, John L. *Wine: A Cultural History*. Chicago: University of Chicago Press, 2011.

Verster, Joris C. "The 'Hair of the Dog': A Useful Hangover Remedy or a Predictor of Future Problem Drinking?" *Current Drug Abuse Reviews* 2, no. 1 (2009): 1–4.

Watkins, Nikki. "So Hungover I Missed My Wedding . . ." *Sun* (London). July 4, 2011.

Wells, David. *Perfect, I'm Not: Boomer on Beer, Brawls, Backaches and Baseball*. New York: William Morrow, 2003.

Wodehouse, P.J. *Ring for Jeeves*. London: Arrow. 2008. First published by Herbert Jenkins in 1953.

Wolfe, Tom. *Bonfire of the Vanities*. New York: Picador, 2008. First published by Farrar, Straus, Giroux in 1987.

———. *The Right Stuff*. New York: Picador, 2008. First published by Farrar, Straus, Giroux in 1979.

Wurdz, Gideon. *The Foolish Dictionary: An Exhausting Work of Reference to Un-certain English Words, Their Origin, Meaning, Legitimate and Illegitimate Use, Confused by a Few Pictures*. Boston: Robinson, Luce, 1904.